INDUSTRIAL SENSORS AND APPLICATIONS FOR
CONDITION MONITORING

About the Author

Phil. Wild was born in Nottinghamshire in 1969. He moved to the seaside town of Scarborough where he attended comprehensive school and sixth form college.

After Sixth form college he started work with Lucas Industries as a trainee engineer. He worked in many different disciplines in sites in Kent, Suffolk, Gloucestershire, and France. Whilst working for Lucas he studied full-time at university for a Manufacturing Engineering degree.

Phil left Lucas in 1992, moving to 3M (UK) plc, where he now works as an optimised engineer.

This book was written in his spare time, based on his experience in industry and his engineering education.

Industrial Sensors and Applications for Condition Monitoring

by

Philip Wild, BEng

Mechanical Engineering Publications Limited
London

First published 1994

©Philip Wild

ISBN 0 85298 902 4

A CIP catalogue record for this book is available from the British Library.

Typeset by The Type Bureau Ltd., Bury St. Edmunds, Suffolk.
Printed by Moreton Hall Press Ltd., Bury St. Edmunds, Suffolk.

PREFACE

The success of the trend towards greater quality and reliability with minimal operator supervision depends to a large degree upon the development and application of automatic sensing techniques. These are required to monitor the performance of machining processes and to compensate for uncertainties and irregularities of the work environment. The benefits of such systems are from increased productivity via the use of extra 'unmanned' shifts, improvement of floor-to-floor cycle times, and reduced scrap levels. Further benefits are obtained by the ability to carry out condition monitoring and predictive maintenance on the complete machine system.

The range of sensing devices and techniques presently available is vast, but only those which are in productive use in industry, and those thought worthy of further development and application are discussed here. These are discussed, where possible, in categories according to the parameter which is sensed and when in the machine process the measurement is made.

CONTENTS

ix

Introduction

WHAT ARE SENSORS?

Definition of a sensor:

'A device to detect, record, or measure a physical property'

We know that human beings have five senses:

- sight;
- hearing;
- smell;
- taste;
- touch.

The functioning of our bodies relies wholly on these five senses, and without any one of them our lives would almost certainly be very different, and we would require special help to compensate for the lack of information received by our body. Our senses are functioning all of the time and are being constantly monitored by our brain to help us in being perceptive, conscious, accurate, appreciative, and to understand.

It is possible to compare the human body (the brain and the senses) to an array of industrial sensors, with a central processor receiving all of the sensory information and interpreting it into physical phenomena.

Thus, each of our human senses can be replaced by single or

multiple arrangements of sensors. These sensors can be linked together to give feedback to each system, allowing each system to be conscious of the other systems. When all of these sensory systems are processed, a perception can evolve of the surrounding environment to which the sensors are being adapted. With each stimulus from the surroundings an impression can be created or changed.

It can thus be seen that a collection of sensors and systems of sensors can be held together to monitor a particular situation in the same way that a normal human being does. Therefore, we can use such systems to monitor:

- machining operations;
- conditions of tools;
- conditions of machines;
- stock control;
- work in progress;
- identification of parts, tools, pallets, etc.

These systems can either be used give a machine a 'hands free' operation, supplementing the operator and giving him or her time to concentrate on other areas of work, or they can be used to complement the operator, helping in the work, providing more accurate data, or taking over the more hazardous jobs.

Instead of allowing a normally open-loop system in machining, they can supply a closed-loop system by taking action to control the machine, depending on the information received by the sensors.

Human beings are sometimes said to have a sixth sense. This is a supposed faculty giving intuitive or extra-sensory knowledge. Some people may claim that they can see into the future with this sense. Well, just as industrial sensors can simulate the five human senses, so they can also simulate the sixth sense.

This is achieved through such means as predictive maintenance, which is a part of condition monitoring. Data are continually taken from the machining system, and these data are analysed to determine the condition of the systems, and the trends of how the conditions have changed since previous analysis. This, together with technical information on parts and machinery, can be used to 'predict' when a machine tool is going to break, when a shaft will be worn excessively, and many more feasible options.

SENSORY SYSTEMS

We can consider improved efficiency, lower supervision, and unmanned machining with these systems. One primary aim is the closed-loop control of the component dimensions by feeding back in-process measurements to the machine controller. The problems posed by the hostile environment in which such measurements may have to be made can be resolved by indirect or delayed measurements.

It is also important to consider the requirements of such sensory systems for their implementation and acceptance by the organizations and individuals involved.

WHEN SENSED

Listed below are points within the manufacturing process at which the component, the tool or other item can be inspected.

(1)   Pre-process.
(2)   In-process.
(3)   Between cycles.
(4)   In-cycle.
(5)   In-cycle measuring station.
(6)   Out-of-cycle.

(1)   Pre-process measurements are made with the component on the machine prior to commencing the machining process, and usually involves component/tool identification and location.

(2)   In-process measurements are made whilst 'cutting metal', and represent the ideal situation.

(3)   Between-cycle measurements effectively utilize any spare time

during machining, for example; one component feature or a tool parameter may be assessed whilst another component feature is being machined or another tool is in use.

(4)   In-cycle measurements include intermittent inspection of partly processed workpieces and final dimensional checks. In each case, machining is stopped to allow inspection of the component.

(5)   In-cycle measuring stations provide location inspection, either on or close to the machine tool, of completely processed workpieces. Components are thus away from the work zone and the information obtained may be fed back to the controller, perhaps with a delay of 1–2 components.

(6)   Out-of-cycle inspection involves transfer, often via a washing station, to a separate inspection facility. This may be a dedicated gauging fixture, a coordinate measuring machine (CMM), or other forms of more traditional inspection.

In this volume we are mainly concerned with:

- in-process;
- between cycles;
- in-cycle;
- in-cycle measuring stations.

ONE

Machine Tool Monitoring System

1.1 INTRODUCTION

Using a machine tool monitoring system it is possible to detect any abnormal process conditions, such as:

- tool wear;
- tool breakages;
- missing workpieces or tools.

In addition, data can be provided on the tool condition and history for the purposes of tool management.

The system avoids unwanted stoppages caused by collisions or tool failure, which leads to a reduction in downtimes and reduced scrapage. In addition, tools will only be changed when it is necessary.

Each tool on the machine is monitored, and just before the cutting forces reach critical values, the monitoring system will take action (if necessary the machine can be stopped in a very short space of time). In this way the operator is relieved of the responsibility of constantly supervising the cutting operations, allowing other tasks to be carried out.

On any one machine up to 18 different sensors can be monitored using a standard process monitor control box; using a personal computer (PC) up to 300 sensors can be monitored simultaneously.

Such monitoring systems can be adapted to a number of machining operations, such as:

- drilling;
- single- and multi-axis turning;
- pressing;
- punching;
- assembly operations;
- horizontal and vertical milling.

When using a PC monitoring up to 300 sensors it is possible to control:

- customized systems such as transfer lines;
- machining centres;
- special machine tools.

There are many techniques for monitoring the tool on a machine, the most common of which are listed below.

Sensors

(1) Feed force sensors:
    (i)   flat plate sensor;
    (ii)   cylindrical sensor.

(2) Current/power sensors.

(3) Press sensors.

(4) Accelerometers.

Integrated systems

(5) Torque Controlled Machining.

(6) Tool setting.

All these systems give an output signal which is indicative of the state of the tool(s) on the machine. This output has to be decoded into a useful form to enable the operator to make decisions, or for an automated function to be carried out. This is achieved by using a modular monitor for process control. Figure 1. shows the system configuration.

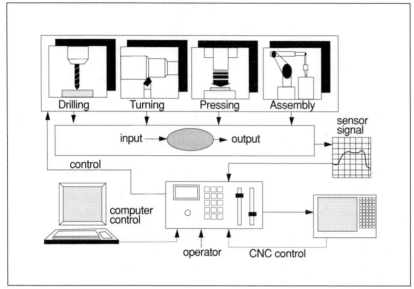

Fig. 1.1   Tool monitoring system

There are two fundamental types of monitoring systems; one is suitable for retrofitting and provides a basic monitoring facility, and the other is for adapted machine parts, and permits high performance monitoring.

1.2   FEED FORCE SENSORS

The force sensor is used for converting the mechanical forces, generated at the tool cutting edge during machining, into electrical signals which can be sent directly to a signal processing instrument.

During the cutting operation a tool will produce relative high loads as it wears. The axial force provides an indication of the condition of the cutting edge (force is a function of tool wear). See Fig. 1.2.

This change in force is instantly received by the feed force sensor. The signal from the sensor is analysed by a signal processing device. This device can then take action, through the machine controller, in

7

Table 1.1

Basic monitoring system (suitable for retrofitting)	High performance monitoring system (for adapted machine parts)
current/power sensors	feed force sensors, plate sensors, and press sensors

– collision detection
– short hole drilling
  (breakage detection)
 	
– HSS and deep hole
  drilling
  (dia. >10mm)
  (breakage detection)
– parting off protection
  (breakage detection)
– rough turning
  (breakage detection)
– missing tool/workpiece
	– collision detection
	– drilling dia. >10mm
	  wear monitoring/breakage
	  protection
	– turning, wear monitoring/breakage
	  protection
	– electronic switch activation
	– missing tool/workpiece
	– pressing operations
	  (assembly status)

case of tool wear, tool breakage, or of tool-not-in-cut.

There are two main types of force sensors; the feed force cylindrical sensor and the flat plate feed force sensor. See Fig 1.3.

1.2.1 Cylindrical sensors

1.2.1.1 Description

These sensors can be fitted in most types of machines. They are principally suited for integration with the bearing packet supporting a rotating shaft or spindle. This may be a machine tool feed screw or a main spindle, depending on its application.

The cylindrical sensor may also be mounted on non-rotating shafts, such as feed bars.

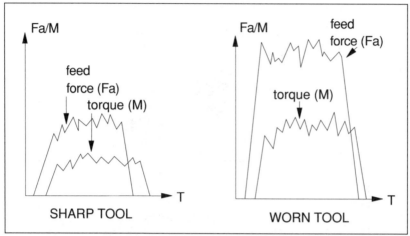

Fig. 1.2 The differences between a sharp tool and a worn one

1.2.1.2 Installation

Sensor installation is optimal in new machines, but in certain cases, retrofitting can be undertaken. The sensors are separate units and can be connected directly to a tool monitoring system.

The sensor is installed as a bushing between the axial bearing and the machine housing. Mounted in the feed-screw or main spindle of the machine tool, the sensor permits the axial force to be measured, and monitors almost directly. The short transmission paths of the forces help the reaction times to be kept to a minimum.

1.2.1.3 Applications

The sensors can be adapted to most types of machine tools. Typical examples are as follows:

– drilling machines, either single or multi-purpose machines, such as stand-alone units and transfer lines,
– NC turning machines,
– NC milling centres,
– special purpose machines.

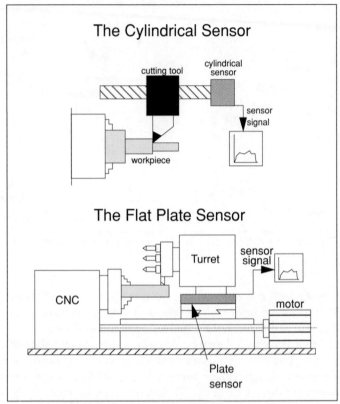

**Fig. 1.3   Schematic views of tool monitoring systems using
feed force sensors**

Feed screw applications

Feed force sensors are installed in the axis of the feed drive's bearings
(any of the x, y, or z axes). An example of the force rating for a typical
NC lathe would be 40kN, although this nominal force will depend on
the MTD or end-user requirements. The sensors can be designed for
angular as well as taper roller bearings, or for combined axial and
radial bearings, and so on.

Main spindle application
In multi-spindle drilling applications, where a common feed slide is used, sensors should be mounted on individual spindles to attain an accurate indication of wear on each cutting tool. A typical nominal force rating, for example, on a six-spindle deep hole drilling station, would have a rating of 1.2kN.

Non-rotating feed bar application
Sensors can be mounted in a non-rotating feed mechanism, where no bearings are used. An example of such a machine would be a six-spindle chucking auto. Sensors with a force rating around 20kN would be used in this application.

1.2.1.4 Feed force sensor selection

Most sensors are individually designed for specific requirements, but three basic steps can be followed to determine the requirements.

(1) Decide where the sensor should be installed in the machine (e.g., main spindle), but note one end of the shaft should be free to allow the forces generated to flow without interference.
(2) Determine the type of bearing that is to be used, and the space available.
(3) Sensor prices for this type of sensor are determined primarily by their diameter.

1.2.2 Flat plate sensors

1.2.2.1 Description

As the name suggests, this sensor is a flat rectangular sensor which, with speed and accuracy, measures the cutting tool condition during metalcutting operations. The plate is installed close to the cutting edge, directly in the path of the transmitted feed force (see Fig. 1.3).

The sensor ensures that, if the machine tool crashes or breaks, no damage is inflicted on the machine. This means reduced downtimes, optimum tool life and minimum scrap caused by tool wear and breakages.

Table 1.2   Cylindrical feed force sensor specifications

Nominal forces:	1–100kN (Special applications can exceed 100kN)	*Linearity*: ±1%
		Supply voltage: 10V d.c.
		Bridge resistance: 360Ω
		Mechanical over-load range:
Measuring device:	Strain gauges arranged in a full Wheatstone bridge with four accurate branches.	800 % of nominal force
Nominal sensitivity:	0.3–2mV/V	

The plate sensor consists of several sensitive zones, each containing complex strain gauge configurations. The design and positioning of these zones provide optimum feed force measurement.

The design ensures that the most critical forces on all axes are measured, and unwanted forces are eliminated.

1.2.2.2   Installation

Thanks to the simple design the flat plate sensor can be fitted to most turning, drilling, or special purpose machines, with minimum cost and engineering changes. For example, in a turning machine a single plate sensor can be installed between the turret and cross-slide to provide monitoring of all the cutting tool axes. This is shown in Fig. 1.3.

1.2.2.3   Applications

The plate sensor similar to the cylindrical sensor can be adapted to most types of machines such as turning, drilling and other special metalcutting machines, but unlike the cylindrical sensor the plate sensors' installation is far simpler as there are no major engineering changes that are required. Note the basic plate sensor can be customized to suit the exact requirements of the end user.

Turning machine applications
The plate sensor is designed and positioned so that all the forces acting in the x and z axes are transmitted without loss. The sensors are easily installed between the toolholders and the main slide. (Note: for vertical turning machines the sensor is installed behind the turret.)

Drilling and special purpose machine applications
In this application the plate sensor is fitted between the drilling spindle and the bedslide. In certain applications the plate sensor can also be installed between the workholding fixture and the machine table.

1.3 CURRENT/POWER SENSORS

1.3.1 Description

The function of this sensor is to detect changes in cutting tool condition during a machining operation, by sensing changes in the electrical current supplied to the feed or spindle drive motors or

Table 1.3 Flat plate feed force sensor specifications

Nominal forces:	1–100kN (Special applications can exceed 100kN)	*Linearity*: ±1% *Supply voltage*: 10/24V *Bridge resistance*: 700Ω *Mechanical over-load range*:
Measuring device:	Strain gauges arranged in a full Wheatstone bridge with four accurate branches.	800% of nominal force
Nominal sensitivity:	0.1–1mV/V	

servos.

The current sensor is a simple, compact, and reliable device which can easily be installed into almost all types of electrically-driven machine tools, without any special engineering design.

The sensor is used in conjunction with a tool monitoring control unit to provide a comparatively low-cost, yet effective, tool monitoring system in new or existing machines (see Fig. 1.4).

When the cutting edge of a tool becomes worn or broken, a change in power consumed by the feed motor normally occurs. This change is detected by the current sensor and a signal is sent to the tool monitor control unit for processing and initiation of machine control action.

A single conductor from the power cable supplying the feed motor is fed through the current sensing ring (one or more turns). The current carried by the conductor passing through the sensor ring is transformed into a proportional voltage signal. This voltage results in a sensor output signal which is proportional to the feed force.

1.3.2 Installation

The current sensor is connected to a conductor from the power cable supplying the feed motor. One or more turns of the conductor from the motor may be fed through the sensor, depending on the current rating and change expected in the motor (see Fig. 1.5):

Two windings for 37.5A;
Three windings for 25A;
Five windings for 15A;
Seven windings for 10.7A;
Ten windings for 7.5A;
Fifteen windings for 5A, and so on.

Thanks to such a simple design this sensor can be adapted to almost all electrically driven machine tools.

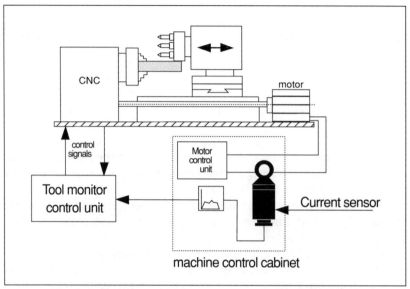

Fig. 1.4   The current/power sensor; sensor configuration

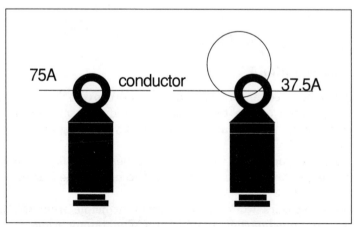

Fig. 1.5   The current/power sensor; conductor turns

1.3.3 Applications

The current sensor can be used in almost all metal-working machines.

Turning.
Milling.
Drilling.
Grinding.
Sawing.

Existing machines can quickly, and at low cost, be fitted with tool monitoring systems, with the sensor operating with a.c. or d.c. feed on main spindle drive motors and servos.

Only one conductor is required to be fed from the motor or servo through the sensor, and the sensor is compact enough to be installed into a control cabinet.

For drilling and turning applications, monitoring the feed forces will normally provide a good indication of the tool condition. It is also possible to monitor the main spindle motor. In this case the resulting signal will indicate the torque exerted on the tool. This, however, is not necessarily the most significant parameter when monitoring the tool condition.

1.4   PRESSURE SENSORS

1.4.1   Description

Similar to the force sensors, pressure sensors convert the mechanical forces generated at the tool cutting edge during machining into electrical signals which can be sent directly to a signal processing instrument.

Pressure sensors, or, as they are sometimes called, force transducers, are rugged devices, yet are capable of accurately measuring forces as low as ±10g.

Pressure sensors work on a different principle from the force sensors mentioned earlier. Instead of a Wheatstone bridge system,

Table 1.4   Current/power sensor specifications

Current measuring range:	0–75A a.c. and d.c.
Operating frequency:	50–60 Hz for a.c. motors
Sensor response time:	0.1 ms d.c.
	25.0 ms a.c.
Linearity:	1% of full scale over a temperature range of 0–50°C
Suitable for use with:	Single-phase a.c. motors
	Three-phase a.c. motors
	Transistor d.c. servos
	Thyrister d.c. servos
Supply voltage:	10v d.c. (20mA)
Temperature range:	0–50°C
Humidity:	10–90% RH
Nominal output for d.c.	2.5V d.c. at 75A a.c.
Nominal output for a.c.	2.5V d.c. at 75A a.c.
Electrical protection rating:	IP65

the pressure sensor contains an LVDT (linear variable differential transformer) the core of which is directly coupled to two conservatively-stressed elastic elements. On loading, the elastic elements deflect the LVDT, resulting in the production of a signal directly proportional to the axial load.

Compared to other force sensing devices the pressure sensor has a greater performance in accuracy, sensitivity, and repeatability (owing to the design of the frictionless operation). Also, this sensor has the capability of detecting accurately both compressive and tensile forces (tensile forces may be present in many machining operations, especially turning).

1.4.2   Installation

These sensors are, optimally, integrated into the machine, but in certain cases the sensors can be adapted into a retrofitting device. The pressure sensor is positioned in the machine so as to receive the axial forces from the machine tool.

The pressure sensor can actually be positioned on a machine tool, acting as an extension to the tool itself (see Fig. 1.6), with the aid of special adaptors. The sensor is held in the tool holder. Other possible installation areas include between the slide bed of the machine and the turret on turning machines, and on workstations, special machining centres, drilling, and milling machines the pressure sensor may even be positioned under the workpiece (see Fig. 1.7).

1.4.3 Applications

This type of sensor is so versatile and unspecialized that the applications for their use are almost unlimited, as they can be adapted to almost all types of machine tool.

Both a.c. and d.c. operated sensors are available to meet particular requirements: a.c. operated units utilize external signal conditioning; d.c. operated sensors incorporate all the necessary electronics.

Various adaptors and load fittings are designed for use with pressure sensors. Retaining nuts, couplings, and mounted flanges are available to hold the sensors in position. Also, the sensors can have a variety of heads, such as hooks, eyes, swivel bearings, load platforms, taper points, and buttons.

1.5 ACCELEROMETERS

1.5.1 Description

As the name of this sensor suggests, it is used in detecting vector acceleration, with very high accuracy. The accelerometer can be used to detect vibrations generated at a machine tool, as both the form of vibration and its amplitude dictate the working condition of the tool. There are many types of accelerometers on the market presently, the most common of which include the following.

Piezo-electric.
Solid state (IC based).
Silicon micro.
Linear servo.
Linear open-loop.

Table 1.5 Pressure sensor specifications

Feature	a.c. operated	d.c. operated
Linearity	>0.2% full range	>0.2% full range
Repeatability	>0.1% full range	>0.1% full range
Operating temperature	−65°F−+200°F	+25°F to +200°F
Survival temperature	−65°F−+300°F	+25°F to +140°F
Excitation	1–5V rms 400–10K Hz	±15V d.c. at 15 mA
Load impedance	400KΩ	2KΩ

Each type of accelerometer has its own characteristics and is suited to a particular application, with its own operating style. Some of the accelerometers have dual applications; as well as measuring vibration they are able to act as inclinometers or tilt sensors.

The sizes of accelerometers vary considerably, from a tiny 3mm package solid state type to a 150mm linear open-loop type, and so the

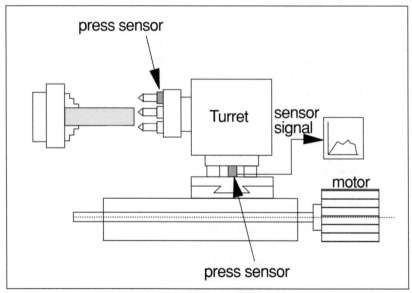

Fig. 1.6 Pressure sensor

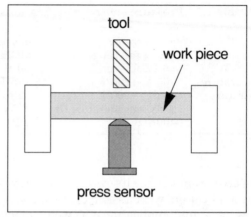

Fig. 1.7   Pressure sensor

installation for the different types of sensors also vary accordingly.

1.5.2   Installation

It is essential that the accelerometer receives the correct vibrations from the machine. That is, the vibrations generated from the machine tool on the workpiece must not be interfered with by vibrations generated from gears or motors. Some weak vibrations may be filtered out in processing.

Owing to this restraint, the accelerometer must be positioned to give optimum reception of the vibrations, without causing a hindrance to normal working. Therefore, it may be positioned on the tool post (for larger accelerometers), or on the tool itself (for smaller accelerometers).

1.5.3   Applications

The accelerometer has many applications, such as in guidance systems for torpedoes, satellites, and fault detection on high speed railways, but this device can also be used to measure frequency vibration in machine tool monitoring.

As the accelerometer is so accurate it can not only determine the state of the tool when in its working condition, but by analysing the form of the vibration it can show if the individual cutting faces on the tool are functioning correctly (see Fig. 1.8). By analysing the tool over a period of time the accelerometer can determine the wear rate of the tool in different working conditions. This may be used in determining optimum settings and setting tolerances for the tool.

1.6 TORQUE-CONTROLLED MACHINING (TCM)

1.6.1 System description

Torque-controlled machining (TCM) is a microprocessor-based control system which can be tailored for adaption to most machining systems. TCM senses the spindle drive motor, armature voltage and current, and spindle r/min. It automatically calculates the tare torque (the torque to cut through air) each time a new spindle speed is selected and eliminates possible variations during warm-up or due to different speed ranges.

The system covers most machining conditions and increases 'up time' and production by permitting torque limits to be programmed accurately at the cutting tool.

As machining begins, the tare torque is subtracted from the total torque acting on the drive motor. TCM then compares this net torque to the programmed torque limits. If it exceeds the values indicated by the programmed code, an appropriate action is taken, based upon the limit exceeded. Consequently, feedrate is reduced, coolant is turned on, or the cycle is interrupted. Also, once the horsepower exceeds the rated machine power or programmed limit, the feedrate is progressively reduced to constrain the power consumed.

1.6.2 System benefits

1.6.2.1 Automatic broken tool protection

TCM can be programmed to detect cutting torque within a specified

21

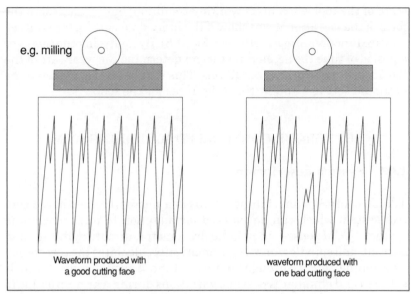

Fig. 1.8   Accelerometer vibration detection

Table 1.6   General specifications for all models

Standard acceleration ranges:	±1.0, 3.0, 5.0, 10, 30g
Acceleration limit:	A steady acceleration of 5 times rated range on any axis will not cause any permanent change in specified performance.
Electrical excitation:	10V a.c. (rms) nominal
d.c. coil resistance:	750Ω/coil nominal
Coil inductance:	120 mH/coil nominal
Carrier frequency:	5kHz nominal
Output:	≥25mV/V at 5KHz for ±1g
Full range output:	50mV/V at 5KHz for 3–30g
a.c. residual imbalance:	≤+/– FRO at 25°C

feed distance. If a contact is not made, no torque will be generated, indicating that either the tool is broken or the part is missing. A cycle interrupt will then be initiated.

1.6.2.2   Automatic coolant control

TCM controls coolant 'on' and 'off' through tool torque sensing as the tool enters or exits the workpiece.

1.6.2.3   Machine and tool protection

TCM will detect when the machine or the tools are overloaded, withdrawing the tool from the work before damage occurs. Spindle reverse and tap withdrawal will operate automatically to provide an especially useful feature when tapping.

1.6.2.4   Other benefits

Greater tool precision is achieved through a combination of automatically adjusting the feedrate to give an optimum value, and programming a torque value to protect a tool from breakage.

Quicker set-up times can be achieved through the use of programmed codes. The tool in the spindle can be automatically moved to touch the surface of the workpiece. The system will detect the surface of the component and update fixture offsets in the same manner as a surface sensing probe. If the surface is to be machined, then the need for a sensing probe is eliminated.

Less time is wasted in cutting air, as the system will automatically increase the feedrate over the programmed air gap. Also, in machining composite materials the system will automatically change the feedrate when encountering the second material.

1.7   TOOL SETTING SYSTEM

1.7.1   System description

The tools sensor setting feature permits tool length setting on a machine. This is achieved in a fraction of the time taken by manual methods. It is to be used in conjunction with a machine table, so with the machine tables movements the tool lengths can be measured and set automatically, without operator intervention, once tools are loaded in a storage magazine.

X and *Y* dimensions can be programmed to introduce an offset when required; for example, for milling cutters or boring tools.

This system cannot be used with machines which do not have a machine table. This constraint determines which machines may be used with this system.

1.7.2   System benefits

1.7.2.1   Tool length checking and re-setting

This system permits the checking of tool lengths at any time determined by the programmer. The tool can be programmed to make contact with the tool sensing device at any time that checking or, if necessary, re-checking of the tool length is required, re-establishing the tool length. For example, prior to machining close tolerance features.

1.7.2.2   Tool length setting

The sensor approaches the tool at a rapid rate. On contact, the sensor retracts and re-approaches at a setting rate. The tool moves in its holder to allow a set length to be achieved. The length is stored along with the *x* and *y* offsets.

1.7.2.3   Broken tool detection

The programmer can include a tolerance in the tool checking cycle which may be used to detect a significant variation from the previously established tool length. If this tolerance is exceeded, a warning message will be displayed on the control screen. The machine will then stop and the operator will be alerted to check the tool.

On fully automatic machines, or in a flexible machining manufacturing system (FMS), both tool and workpiece can be changed to await examination.

TWO

Identification Systems

2.1 INTRODUCTION

Identification systems can be adapted to a multitude of fields in the industrial environment. It is essential for full control in:

- Process planning and control;
- inventory and stock control;
- MRPI and MRPII.

The system has the ability to determine types of parts, machines and machine tools, pallets, and so on. Information, such as the numbers of parts there are and their position, both in the factory and on the machine, can be determined, and controlling action can be taken to contain such parameters as above.

Identification sensors can come in many different forms. The following are the most common types used in industry.

(1) Probing systems.
(2) Read–write chip.
(3) Proximity sensors.

2.2 PROBING SYSTEMS

2.2.1 Introduction

Probing systems evolved in the industrial workplace when Rolls–Royce first started building the engines for Concord. From this origin probes have become extremely accurate and are essential in many machine tools. Other names for probes include the following.

> Touch sensors.
> Trigger probes.
> Surface sensors.
> Coordinate measuring devices.

They are multi-directional devices which are used primarily to determine dimensions from workpieces. They are electronically coupled to the control system of the machine tool. Therefore, probes can be used to determine the following.

> Workpiece present/missing.
> Correct workpiece identification.
> Features machined correctly.
> Angle of workpiece.
> Locate the part programme to zero (align the tool).
> Detect and compensate for core shifts.
> Align the machine to previously machined surfaces.
> Automatically compensate for tool wear.

Probes provide a delicate sense of touch on a workpiece. As the probe is deflected by contact with the workpiece an instantaneous response, which is determined by the magnitude of the deflection, is signalled to a controller. This information is then used to modify the path of a CNC programme.

The probe is mounted on a quill, usually by means of a probe head. Heads may be manually or automatically orientated to achieve the required inspection angle.

Triggering is preceded by limited pre-travel as the stylus deflects. Pre-travel is related to stylus length and gauging force, and it is automatically compensated for during stylus tip calibration.

The life time, reliability, and repeatability are generally very good with most of today's probing systems.

The majority of applications for probing systems are on CNC machine tools, as the control generated from the probing signal is easily utilized in a CNC system. There are many different types of probes on the market today these can be generalized into two categories.

(1) Contact probes.
(2) Non-contact probes.

2.2.2 Contact probes

Contact probes, as previously described, are mechanical devices

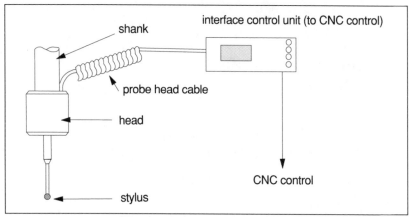

Fig. 2.1 Contact probing device

which rely on physical contact to produce a change within the sensor, which in turn sends a signal to a control monitoring device.

The main elements of a contact probing device are shown in Fig. 2.1.

The main sensing device is the head. This is where the signal is generated (in the x, y, and z directions) giving an output to a control device.

Probe heads perform a dual function; they provide a means of mounting the probe onto the machine quill, and, in most cases, they can be manually or automatically oriented to point the stylus in line with the inspection requirements.

The contact section, the stylus, is the member which physically makes contact with the workpiece. It is made up (usually) of a ruby ball which gives maximum accuracy for most applications (see Fig. 2.2).

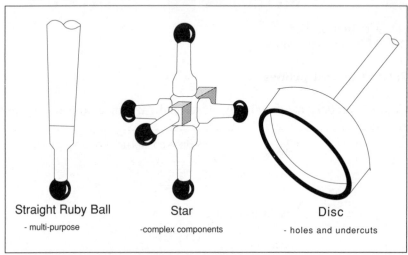

Straight Ruby Ball	Star	Disc
- multi-purpose	-complex components	- holes and undercuts

Fig. 2.2   Probe styli

There are, however, four main categories of stylus:

- ruby ball;
- star;
- disc;
- special styli.

The ruby ball, as mentioned, is the most common and is used in most applications. It has a highly spherical industrial ruby ball bonded onto its non-magnetic stem.

Star styli are ideal for inspecting complex components. They incorporate four or five ruby balls on stainless steel stems, all at 90 degrees to each other.

30

Disc styli are used to inspect holes and undercuts. Special styli can be adapted to measure:

- thread inspection cylinders;
- pointers to inspection points, scribed lines, and thread depths;
- ceramic 'hollow balls' for deep feature probing and rough surfaces.

To connect the probe head to the quill of the machine a shank is used. The correct shank must be chosen to suit the monitoring facility of the machine.

There are a wide variety of different types of heads, styli and shanks, each for its own particular application.

An autochange system is available, for use in conjunction with probing devices. Probes can be automatically changed in seconds to measure different features on a component.

2.2.2.1 *Retrofitting*

For use on a wide range of turning machines, a probe with optical transmission is available. This does not have any connecting cables. It is clamped to the machine turret and is driven against the workpiece. On each occasion when contact is made with the surface, a signal is transmitted to an optical sensing device which, in turn, is connected to an interface and a control monitoring device.

2.2.2.2 *Scanning probes*

The whole surface of a workpiece can be analysed using a scanning probe. A picture of the profile of the workpiece can be created at high speed and with the same accuracy as is provided by conventional probing devices. Scanning can be achieved from both vertical and horizontal mountings, making it suitable for five axis machines.

2.2.3 Non-contact probes

This probe uses a visible spot laser providing a non-contact solution to the measuring of both rigid and soft engineering materials. Data collection is fast, producing up to 200 accurate points per second.

The probe operates at 50mm above the workpiece with a measuring range of ±5mm about the focus point. As the same basic components can be adapted to both contact and non-contact probes, the two systems can be easily interchanged.

The system can be used as a scanner for surface analysis, providing accurate dimensions from the surface, with no need for tip diameter compensation.

The laser probe uses the widely adopted principle of triangulation to rapidly measure variations in distance between the probe and the workpiece. Low power visible laser light is focused onto the surface under inspection and is collected by high accuracy optical elements. The light is then focused onto an analogue position-sensitive detector, which provides a means of measuring the distance from the probe to the workpiece. Automatic compensation for variations in surface reflection enables a large range of engineering materials to be measured.

Non-contact probes can be adapted to suit the same machines as contact probes.

Table 2.1   Probe specifications – contact probes

Principle application:		From low cost manual to very high accuracy, with all styli lengths
Sense directions:		Omni-directional ±x, y, and z
Uni-directional repeatability:		0.25–1.0µm
Pre-travel variation:		±0.15–±2.0µm
Weight:		22–128g
Stylus force range:		7–50g
Stylus over-travel:	x, y axes:	±14 °–±22 °
	z axis; +z	4–10mm
	−z	1.5mm
Max. extension:		100–300mm
Mounting options:		Shank to suit thread or autojoint

Table 2.2 Probe specifications – non-contact probes

The Laser	
Max. output power:	5mW
Emitted wavelength:	830nm
Approx. spot size:	25μm diameter
Nominal Ocular	
Hazard Distance	
(NOHD):	160mm
The System	
Stand-off:	20mm
Measuring range:	4mm (±2mm)
Position sensing range:	6mm (±3mm)
Resolution:	1μm
Repeatability:	2μm
Communication:	RS232
Dynamic performance:	50 readings
Angle to surface:	0–±85 °

2.3 IDENTIFICATION SYSTEMS USING A READ–WRITE CHIP

2.3.1 Introduction

The read–write chip is an electronic tag which acts as a code or data carrier. This enables:

- identification;
- instructions;
- measurements.

It also facilitates the recording of the history of the equipment. Data can be exchanged, inserted, updated, or read anywhere in the work-shop by an adapted PC or machine controller. Data is inductively transferred between the tag and the reading head.

There are two different systems available.

(1)   The read–only system. This uses code carrier tags which contain pre-set codes which must be identified by a controller.

(2)   The read–write system. This uses data carrier tags which store data which can be updated and exchanged with a controller.

2.3.2   Read–only system (see Fig. 2.3)

The read–only system utilizes tags containing a fixed multi-digit code. The system operates with a central processor which stores and processes the data related to the item identity code.

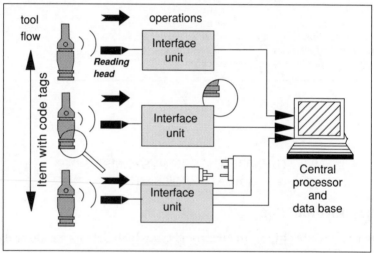

Fig. 2.3   Read–only identification system

The read–only system primarily consists of:
–   a read–only code tag installed in each item;
–   a reading head and interface unit at each reading position, as well as a suitable software package.

The system operates using a PC or mini-computer as a central processor. This supports reading, communication, database, and processing, together with management procedures.

2.3.3 Read–write system (see Fig. 2.4)

The read–write data carrier tag can typically store 2KB of data. Existing data can be updated and new data entered as well as displayed or transferred.

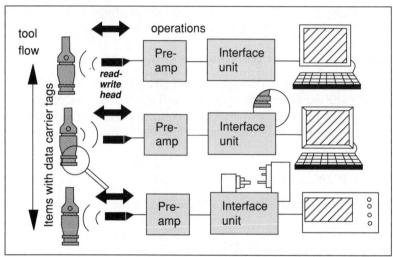

Fig. 2.4 Read–write identification system

This system utilizes independent read–write stations, which can be located anywhere in the factory.

To operate in this configuration, the station requires read–write data carrier tags, a read–write head with a pre-amplifier, and an interface which communicates with PCs and machine controllers.

The electronic tags can be installed in:

- tools;
- fixtures;
- pallets;
- accessories.

2.3.3.1   Benefits of the read–write chip

(1)   Provides accurate tool data flow between tool, operator, and machine as well as tool management.
(2)   Contributes to tool stock rationalization.
(3)   Guarantees data exchange accuracy. Electronic data exchange eliminates operator reading and writing time, and errors.
(4)   Secures set-up and tool changing in the machine.
(5)   Simple installation, with no precision alignment necessary between the tag and the reader.

2.3.4   Tool (and fixture) identification

Tool identification systems are often desirable, to prevent data-entry errors by the machine operator or to speed up the tool-load process and increase operator-utilization. These concepts are valid on standard machining centre applications or on flexible manufacturing systems (FMS).

2.3.4.1   Chip mounting

The chip is inserted into a hole machined in the taper-shank of the tool-adaptor.

2.3.4.2   Chip read–write unit

To read or write tool data the tool adaptor is inserted into a fixture containing the read–write head. This procedure ensures correct alignment of the tool data chip and read–write head, and prevents the possibility of error.

On standard FMS applications, without automatic tool delivery, a read–write unit is mounted adjacent to the tool load point at the rear of each machine.

An interface connects the read–write head to the machine's CNC.

On FMS applications using automatic tool delivery, two read–write heads are used.

(1) At the operator load point. The read–write head is mounted in the load arm. Entry or removal of the tool from the system is recorded.

(2) On the gantry vehicle. The read–write head is mounted on the gantry robot arm. The tool identification code is read on transfer of the tool to/from the machining centre storage matrix, to confirm the transfer.

One interface can be provided, with two connections, one for each read–write head. The interface links to the vehicle controller.

2.3.4.3 Tool data transfer

The route for tool data transfer depends on the configuration of the machining centres.

2.3.4.4 Optical tool pre-setting machine

Tool data is created at the tool pre-setter and is temporarily stored in the tool pre-setter's CNC.

A read–write unit can be provided at the tool pre-setter.

The method of tool data transfer depends on the type of system controller.

2.3.4.5 Flexible Manufacturing Systems (FMS)

On FMS, with automatic tool delivery, the data is transferred via a serial link directly to the cell control computer, and recorded in a non-volatile memory. The tool data is transferred from the cell controller to the machining centre CNC, via the local area network, and then the tool is loaded into the machining centre, either manually or automatically.

2.3.4.6 Standalone machining centres (see Fig. 2.5)

Tools are conveyed manually to the machining centres. The tool load and unload processes are manually controlled.

For this system the following equipment is required at the tool load point of the machining centre.

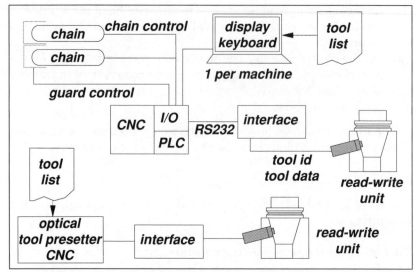

Fig. 2.5 Stand-alone machine or FMS (without auto-tool delivery)

(1) Read–write unit, attached to the machine structure.
(2) Control console, with the following controls:
    (i) numerical display of the tool identification number;
    (ii) numerical keypad, and special function keys.

The operator follows externally generated tool load and unload lists.

2.3.4.7 Tool unload (stand alone m/c centres)

To unload tools, the operator selects the appropriate tool code on the control console. The CNC moves the machining centre storage matrix to bring the selected tool to the unload position. The operator places the tool into the read–write unit, and writes the current tool data into the chip, using the function keys on the control console.

2.3.4.8 Tool load (stand alone m/c centres)

To load tools, the operator selects a tool from the tool trolley and inserts it into the read–write unit.

The tool identification code is read and displayed. Tool data is transferred to the CNC at the same time.

The tool data contains information on the number of tool pockets required to store the tool. The CNC itself selects an appropriate tool storage location and automatically moves the chain to present this location to the tool load point. The operator inserts the tool into the chain to complete the tool load operation.

2.3.4.9 *Tool load/unload on FMS without automatic tool delivery (see Fig. 2.5)*

Tools are conveyed manually to the machining centres. The tool load and unload process is manually controlled.

The following equipment is installed at the tool load point of each machining centre.

(1)   Read-write unit attached to the machine structure.
(2)   Control console with the following controls:
    (i)   numerical display of the tool identification number;
    (ii)   numerical keypad, and special function keys.

The operator follows tools lists generated on the controller.

2.3.4.10 *Tool load/unload on FMS with automatic tool delivery (see Fig. 2.6)*

Tools are conveyed manually from the external tool service area to the tool load/unload point of the automatic tool delivery system.

The following equipment is installed on the tool load point of the automatic tool delivery system.

(1)   Read–write unit, built into the tool load pocket.
(2)   Read-write unit, built into the transfer arm of the gantry robot.
(3)   Control console, with the following controls:
    (i)   numerical display of the tool identification number;
    (ii)   numerical keypad, and special function keys.

The operator follows the tool list generated on the controller.

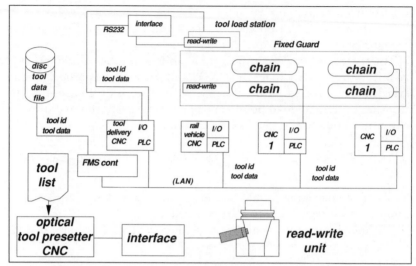

Fig. 2.6   Read–write system with FMS (auto-tool delivery)

2.3.4.11   Alternatives

(1)   Mounting positions.
(2)   Retrofitting devices.
(3)   Sensitivity distances.

2.3.5   Pallet identification (see Fig. 2.7)

The prime benefit of pallet identification is the insurance that pallets actually arrive at their intended destination. This can also be utilized for pallet flow diagnostics.

2.3.5.1   Pallet identification system

The write facility enables the initial assignments of the numbers to the pallets. The number can be changed as required at set-up. During pallet transfer around the factory only the read facility is used.

To provide low cost and maximum simplicity and reliability, a limited number of read–write units are normally used in a system.

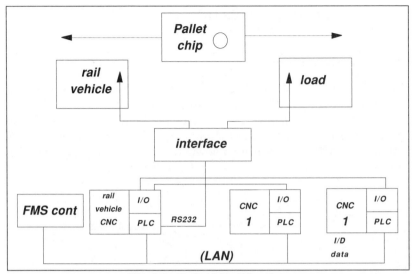

Fig. 2.7 Pallet identification

(a) *Rail-guided vehicle*
One read-write unit is carried on the rail-guided vehicle. The pallet code is read on the transfer to and from the vehicle, for example, at the park stations, at the machining centre pallet-shuttle, or at the load stations.

(b) *Load station*
One read–write unit is mounted at each pallet load station. A pallet identification code can be written into a pallet, if residing on the load station. Alternatively, the system can read the code as necessary.

2.3.5.2 Pallets
Pallets need to be modified to include the implanted read–write chip.

2.4   IDENTIFICATION SYSTEMS USING
PROXIMITY SENSORS

2.4.1   Introduction

Proximity sensors are a crude identification system, but less expensive than the integrated systems, as previously discussed.

The proximity sensor is a retrofitting device which determines the presence of an object within a certain range; for example, counting the number of parts moving on a conveyor system or determining whether a workpiece is present in a machine (see Fig. 2.8).

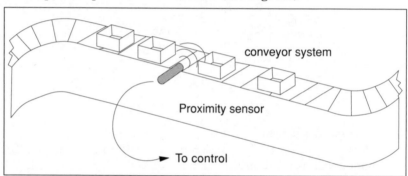

conveyor system

Proximity sensor

To control

Fig. 2.8   A proximity sensor used to control the flow of parts on a conveyor system

2.4.2   Operation

There are three main types of proximity sensors that are commonplace in today's industries.

(1)   Ultrasonic.
(2)   Infra-red.
(3)   Pneumatic.

The ultrasonic and infra-red sensors operate in a similar way; a signal is sent from the sensor (transmitter), and if this signal hits an object, then an echo is created. This echo is picked up by the receiver on the sensor. Therefore, depending on how far away the object is from the sensor, the echo takes a longer or shorter time to return to the sensor. Each sensor can, therefore, be adjusted to only detect echoes from a set distance, so eliminating any 'noise' that may interfere with the sensor, giving a false message.

The infra-red sensor can also detect changes in temperature by detecting discrete changes in amplitude and wavelength of the returning echo.

The pneumatic proximity sensor works in a very different way from the previous two. This sensor does not rely on feedback, in the form of a transmitter and a receiver. It works on the principle that when an object obstructs the path of air, the pressure of the oncoming air is increased. This pressure is monitored to determine the presence of an obstruction. This system has a limited range, and is used in conjunction with a PLC, but the system does have the advantage of being fully compatible with a pneumatic circuit and does not contain any electrical components.

2.4.3 Installation and applications

Proximity sensors have unlimited uses in industry, and some of the most common ones include:

- detection of tools in a tool matrix;
- detection of workpieces in machines/fixtures;
- counting components on a conveyor system;
- sorting components by length on a conveyor system;
- determining presence of people in a hazardous area (safety feature);
- sorting components into batch sizes;
- sorting heat-treated and non-heat-treated parts.

The sensors can be linked into the controllers on machining centres, CNCs, and the like, to take any necessary action – for example, stopping the machine due to the absence of a tool or the workpiece.

Alternatively, such a sensor can work independently of the machine or operation which it is monitoring, by using its own control system – for example, counting components on a conveyor system.

THREE

Non-Contact Vision Systems

3.1 INTRODUCTION

Machine vision systems are gradually being introduced into a wide variety of industries, including, for example:

- food;
- pharmaceuticals;
- automobile;
- steel;
- armaments;
- aircraft;
- electrical.

Non-contact sensing may be invaluable in certain areas of these industries; for example, where the component is very sensitive to contact or where the product is held in hazardous or dangerous conditions.

When using vision systems we must take into account many parameters that are associated with it, including:

- mechanical handling;
- optics and, possibly, fibre optics
- lighting;
- sensor technology;

- electronics (both analogue and digital);
- software;
- algorithms;
- production engineering (including existing QA practices).

Of course, not all vision systems will contain all of the features above, and there may be variations on these to suit differing types of applications.

There may be alternatives to conventional vision systems available that may be used for different non-contact applications.

- ultrasonics;
- microwaves;
- X-rays;
- nuclear magnetic resonance.

There are many differing vision systems presently used in industry today, each varying in complexity and application (see Fig. 3.1 for a typical application).

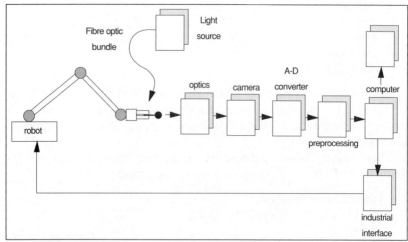

Fig. 3.1 Machine vision system

(1) *Black and white video camera*
Used to analyse and monitor a wide variety of features:

- identification;
- positioning;
- quality – cracks, finish;
- machine and tool condition.

(2) *Colour video camera*

As above, but allowing for colour differentiation.

(3) *Laser systems*

Can be used for the following:
- dimensional analysis;
- size and shape analysis;
- vibration analysis;
- positioning.

(4) *Simple switches (light beam or laser)*

Can be used for the following:

- counting components;
- safety devices on machine covers;
- presence of objects in a machine environment;
- proximity devices.

Each of the above vision systems will be discussed in more detail below.

3.2   TECHNOLOGY AND INFORMATION PROCESSING

There are today two main categories of optical sensors which are available of industrial use:

- cameras with Vidicon tubes;
- optical sensors with semi-conductor targets (CCD).

TV cameras have quite a few serious disadvantages when compared with semi-conductor sensors. Long-term drift is hardly, if ever, compensatable. Analogue tube sensors are very easily influenced by local factors which are either a function of temperature and/or the influence of electromagnetic fields. Such systems are also very sensitive to shock. Conversion of picture data into standard video signals has advantages with respect to being able to use a normal video monitor (e.g., TV screen) but at the same time has disadvantages with respect to the time factor, restricting its use for industrial purposes.

Optical semi-conductor sensors are available either as one dimensional or two dimensional array elements. Data readout by such elements is by one picture element after another. This means there is a direct relationship between a certain point in the picture and a certain point on the semi-conductor target. During this process a variable readout frequency can be used. Apart from the minimal weight and the not-too-complicated electronics, there are other factors, such as linearity and radiation intensity, which have an influence on the choice of semi-conductor sensors.

The resolution of TV camera systems today is relatively low. For a good resolution an image of the order of about 1000 x 1000 elements with an 8-bit grey scale would be used, requiring a memory of 1 megabyte. Such memories are not feasible due to the very high costs. Therefore, various methods to reduce the amount of picture information are used to bring the usage down to a practical level.

(1)   The first method is to use a binary picture by setting a predetermined gray value as the threshold of black and white.

(2)   Secondly, when handling measured values it is often sufficient to count the number of black and/or white elements.

Both these methods of data reduction can be accomplished through software programmes. However, there is also the possibility that certain operations can be quickly and easily taken over by hardware configurations.

Figure 3.2 shows a complete optical sensor system consisting of the following elements:

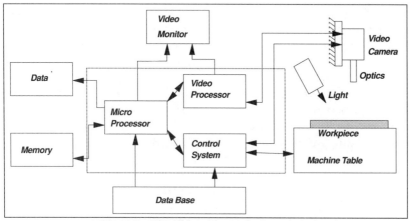

Fig. 3.2 Video camera analysis system

- camera with lens, microcomputer interface, or a preprocessor;
- microcomputer and peripherals for controlling the whole system;
- x–y positioning table.

3.2.1 Application examples

In most optical sensory systems the main problem is the determination of the geometric values. Length, diameter, inner and outer dimensions of two-dimensional objects can be completely inspected. Inspection of completeness during machining and assembly processes can be fully automated using such systems. It is possible to check for attributive characteristics such as missing screws, washers, and assembly parts. As far as attributes can be recognized, it is possible to determine them in the various stages of the manufacturing process or in production.

A special type of problem is the inspection of form, where it is necessary to check the difference between an actual and a given curve, for example, roundness, flatness, cylindricity, or straightness. Here, optical sensor systems have an advantage over conventional testing devices due to the fact that a large number of points on the curve can be laid through the picture grid and can be used very quickly for further processing.

49

The field of surface checking has been a typical problem for visual inspection, previously only being carried out manually. Characteristics being checked are scratches, grooves, burrs, dirt, foreign particles, and contrast differences of surfaces. Those operations can now be achieved by optical sensors.

Features located at a certain angle to the surface of the test components can also be measured with the aid of appropriate illumination techniques; for example the inspection of the burrs produced when stamping sheet metal components.

3.3 VIDEO CAMERA

3.3.1 Inspection using a video camera (crack detection)

Many operations can produce cracks or unwanted notches or scratches. These defects are traditionally inspected manually, but this obviously causes disadvantages in an automated system. If, for example, an industrial robot is being used to manipulate the component, then the component (after magnetizing and spraying with a fluorescent liquid containing magnetic particles) can be placed in a small box containing ultra violet lamps. The robot rotates the component and its surface is scanned line by line by the camera. The binary graphics processing is performed by computer. Points of light caused by dust particles, remains of the magnetic particles, and reflections are 'faded-out' by the computer. Detected cracks are displayed on a monitor and can be automatically assessed by using pre-programmed tolerance limits; the operator/supervisor can be notified, and the component stored as scrap.

This type of system can also be adapted to other applications, such as inspection of a machined part to see, for example, if all holes have been drilled correctly and in the correct position.

3.3.2 Image processing for assembly

3.3.2.1 The image processor

Vision systems have been increasing in automatic machines and robots. Vision systems expand the limits of automation. One of the most important points about vision systems is the development of a low cost and general purpose image processor.

The basic features of the image processor (PPI–1) are as follows.

(1) Recognition algorithms are processed by software.

(2) The vertical and horizontal resolutions of a picture are independently changeable according to the purpose of the application.

(3) To save the image memories the processor (PPI–1) can take a partial picture by window processing.

(4) By computer command, the processor (PPI–1) memorizes a picture in binary mode or grey scale mode (8 bits). This switching

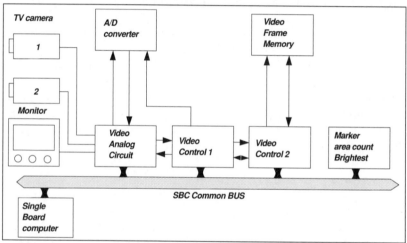

Fig. 3.3 Image processing (PPI–1)

is performed in real time.

(5) The processor (PPI–1) has an arithmetic LSI (APU) which facilitates high calculation rates.

Figure 3.3 shows the image processor (PPI–1). The composite video signal from the TV camera is converted into a binary video signal by a comparator. If the mode is set to grey scale, the composite analogue video signal is converted into an 8-bit video frame memory. Using multiplexing circuits, the refreshes of the image can be switched in real time. The contents of the video frame memory are displayed on the TV monitor.

The choice of optimal resolution and grey level for each application is essential to minimize the processing and analysis time.

3.3.2.2 Assembly

In the assembly of component parts accurate measurement of position is essential. Solid state MOS type cameras can be used as they create no image distortion. More than one camera may be used as this will give more focus points and improve the resolution of the picture (typical resolution 320 x 240, horizontally and vertically, respectively).

Lighting is essential in this process, and usually the most effective form of this is a fluorescent light source.

Once the hardware is set up, then the software is utilized to recognize components and parts. This, in turn, can be used to recognize correctly assembled parts, or to act as a controller for an automated machine or robot. This will eliminate any direct operator control, and permit removal of any mechanical guides and fixtures that were previously used.

3.4   VIDEO GREY SCALES

With black and white vision, to deal capably with an object the background must be of contrasting brightness, so that the object is easily distinguishable.

Instead of processing an essentially binary image, where there are only two grey levels, black and white, multiple levels can be used; for example, up to 256 grey levels, where 0 is black up to 255, which is white.

The processing of grey level images introduces a wide range of extremely powerful techniques which come under the areas of image processing and image analysis. The manner in which they can be applied to industrial problems varies, but the general aim is to divide or 'segment' the image into specific parts (regions or objects) so that a meaningful description can be generated. In order to extract the object or feature from an image, the points that belong to it must be marked. For ease of making the final decision this marking process is usually designed to produce a binary two-valued picture, 1 for points belonging to the object and 0 elsewhere.

With the advent of fast digital image processing techniques the manipulation of grey level images has become acceptably fast, and it is increasingly used in many diverse fields, ranging from medical diagnostics to military surveillance.

This system has obvious advantages in the factory. For example, identification and monitoring of both the parts passing along the conveyor and of the robot arm itself is a considerable benefit.

3.5 COLOUR VISION SYSTEMS

3.5.1 The human eye

Colour vision is probably the only way of distinguishing differently coloured objects with the same shape, size, and features. Colour perception can be considered a very personal experience. It is affected by the physical characteristics of the source and the transmitting medium, by the physiology of the retina and the visual nervous system, and also by the psychological state of the processes of the cerebral cortex in interpreting the signals sent through the retinocortical pathway.

Human colour vision basically treats a colour as a three-part report from the retina. The eye has evolved to see the world in unchanging colours, regardless of shifting, unpredictable, and uneven illumination. This 'insensitivity' to changing lighting makes it attractive for industrial applications.

Colour vision is particularly useful in industrial automation to recognize, search, sort, and manipulate coloured parts or colour-coded objects. The use of colour permits part discrimination where grey scale information alone is insufficient, and often, at times, avoids complicated and time consuming grey scale analysis.

Traditional black and white vision has relied on extraction of silhouettes from the workpiece. The study of machine binary colour vision must deal with acquisition of colours, differentiation of colours, filtering and selective processing of colours, measurement of colours (colorimetry), together with all the features (such as size, shape, position, etc.) that a binary vision system can extract from a scene.

3.5.2 Colour image

A wide range of colours can be reproduced, to the satisfaction of the eye, by the addition of only three monochromatic light sources: red, green, and blue. The three CIE standard primaries are monochromatic light sources of wavelengths 700 nm (red), 546.1 nm (green), and 435.8 nm (blue).

A digital image is defined by a function of two-dimensional position, say $I(m, n)$, defined at chosen grid points of the image. For a chromatic grey scale image, the function I is scalar valued, its value being the brightness of the image at a certain point. For colour images, three values must be specified at each point, i.e., the function I is a vector value and has three components. A common choice of the three components is that of the so called red, green, and blue (R, G, and B) components. The R, G, and B components can be transformed to other quantities, more closely associated with our visual senses of colour, such as brightness, hue, and saturation.

3.5.2.1 Hue, saturation, and chrominance

The hue describes the intrinsic nature of the colour, i.e., red, green, cyan, purple, etc. The colour itself is its hue or tint. A red apple has a red hue; green leaves have a green hue, etc. The colour of any object is determined primarily by its hue. Different hues result from different wavelengths of the light producing the visual sensation in the eye.

Saturation is a measure of colour intensity, i.e., its pastel versus its vivid quantity. Desaturated colours are whitish or washed out, whilst saturated colours are vivid, intense, deep, and strong: i.e., pale or weak colours have little saturation. The saturation indicates how little the colours are diluted by white.

The term 'chrominance' is used to indicate both hue and saturation of a colour. The chrominance includes all the colour information, without the brightness. The chrominance and the brightness together specify the colour information completely.

3.5.3   Colour coding technique

Utilizing colour filters for the three primary colours, colour coding can be performed. Basically, the principle of this operation is similar to

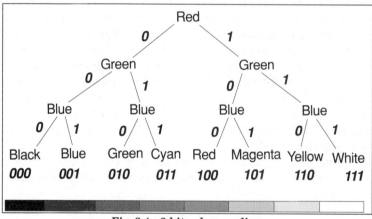

Fig. 3.4   3-bit colour coding

the early design of colour video cameras, which used a rotating colour filter wheel, triggered by the vertical sync. signal, for colour separation.

We can use a three-bit colour coding technique to code and distinguish eight colours. Figure 3.4 shows the binary tree classification method used in colour coding and classification. We can see that a three-bit code works best for the three primaries and their complementary saturated colours. For desaturated colours, because of the dilution by white, ambiguity will arise. e.g., light green, pink, and light blue may all be indistinguishable from white in coding since they are all coded 111. Another example is that yellow and orange may both have the same colour coding, 110; that is, both of them will pass the red and green filter test and fail the blue one, despite that they are different colours as perceived by the human eye.

3.5.4 Conclusions on colour vision

By combining machine vision with colour filters, colours as well as location, shape, and orientation can be easily obtained. This combined colour/location/orientation information can be used by a robot or other manipulator to pick up a distinguished, coloured object. Additionally through the use of filters, irrelevant colours can easily be rejected. This is particularly important in saving computer processing time by filtering out cluttered background noise. Objects with connected multi-coloured regions can be recombined by using other features such as area, centre, and pixel counts. Colour codes can be identified on a complex background. The limitations posed by connectivity analysis can be readily perceived as isolated ones using filters. The same scene will be easily misinterpreted by conventional binary vision system as a single object because of the use of 'blob' processing and run-length encoding.

It is easy to see that the introduction of colour differentials for current machine vision processing can greatly extend the range of applications.

3.6 VISION SENSOR SOFTWARE

The software for sensory subsystems is written mostly in Pascal. The Pascal compiler generates machine instructions, but about 10 per cent of the system is hand coded to decrease the time consumed for image processing. Three levels of representation are used by the sensor algorithms.

(1) The lowest level is the image itself, which is represented as a 256 x 256 array of integers (I) in the range 0–255 (pixels). Pixels are

addressed by a row index and a column index, and the pixel at row (r) and column (c) is denoted $I(r, c)$.

(2) The next level of representation is the 'hit list', which is a 1 x 256 array (H) of row indices in the range 0–255. The ith element of H is denoted $H(i)$. $H(i)$ is defined as the row index where the stripe in the image intersects the ith column of i.

(3) The final level represents an array, s, of segments, where each segment represents a straight line segment lying in I. Components of a segment are the row and column indices of its end points, its length, and its slope. A stripe in the image is represented as an array of segments.

3.6.1 Application

The sensor software provides functions for capturing images of stripes, transforming one representation into another, sorting and comparing segments, and computing the error signals that are fed back to the control computer.

3.6.2 Operation

A robot drives the camera (or the workpiece can move beneath the camera) so that the end-point, or datum, of the workpiece is in the centre of the image. The image is then captured into an array, and the sensor sub-system must now carry out computations and store them in a database with information associated with the given end-points. Inspection of the workpiece can be performed by moving the camera around the workpiece, covering the whole of its area, on each step calculating an array of the image and storing this in the database.

The information in the database can be manipulated to analyse the workpiece.

3.7 LASERS

High power lasers are applied increasingly in various fields of material processing, such as welding, cutting, melting, hardening, and

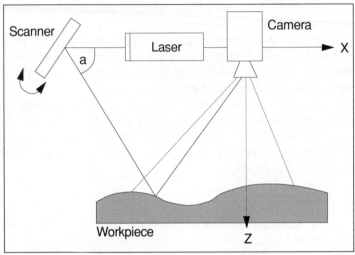

Fig. 3.5 Scanning using a laser

others, but lower powered lasers can be used for less mechanical purposes, for example:

- scanning;
- dimensions;
- surface finish;
- part or feature detection.

Lasers can also be used in conjunction with other vision systems; for example, in Fig. 3.5 a laser is being aimed at the surface of a workpiece and a video camera is determining the scene of the workpiece by analysing the position of the laser.

3.7.1 Laser applications

3.7.1.1 Scene analysis

Laser scene analysers or 'range finders' usually work either on a time-of-flight principle or a triangulation principle. Triangulation has a number of features which make it both easier to use and more flexible

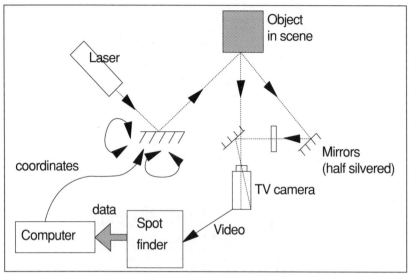

Fig. 3.6   Scene analysis using a laser

in its performance.

Figure 3.6 shows a typical scene analysis system in diagrammatic form. The system is operated by projecting a laser spot through an electro-optical deflection system into the scene. The deflection of the beam is controlled by computer. The scene is viewed by a colour TV camera which has been modified so that signals can be acquired from it directly. There is a special piece of hardware, called the spot finder, in the red channel which delivers to the computer a list of coordinates (in the TV coordinate system) representing the position of any spots present in the TV image, from which the position of the laser spot is calculated. The distance of the spot from the camera is measured by using mirrors to form two images of the spot and then doing a simple triangulation calculation based on the separation of the two spots in the image.

Figures 3.7(a) and 3.7(b) illustrate the ways in which the instruments are set up and used. There is a half-silvered mirror in front of the camera and another, fully-silvered mirror to one side. The mirrors are adjusted for parallelism by pointing the camera–mirror combina-

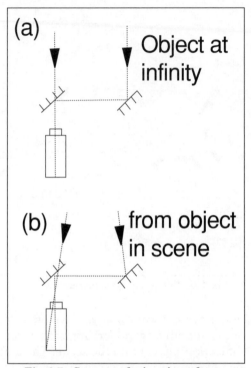

Fig. 3.7 Scene analysis using a laser

tion at a distant object (essentially a parallel light source) and adjusting the mirrors until the two TV images of the distant object coincide (Fig. 3.7(a)).

In use objects are comparatively near and give rise to non-parallel rays through the two mirrors and, hence, to two images on the camera. In this way the two laser spots will be seen and can be used for triangulation (Fig. 3.7(b)).

Using this system we can determine aspects of the viewed object:

- its shape;
- its size;
- its geometric features.

Constraints

However, this system does have some restrictions. The speed at which measurements can be taken may be possible limitations, as the instrument requires about 1 millisecond settling time between measurements. Another, more serious, constraint results from the limitations of the TV system. The photocathode is completely scanned once every frame period (40 milliseconds). If we make only one measurement per frame then we require one or more intervals of 40 milliseconds per measurement. If we try and do more than one measurement per frame, then we have to be careful to avoid the ambiguities caused by having several spot-pairs in the picture. To maximize the measurements the laser scan must be synchronized with the TV scan.

3.7.1.2 Other applications

Lasers are devices that are based on interferometry, with resolutions to 2.5 nm, and so can be used for many applications. Some other, superior, characteristics of lasers may be found to be the best, or even the only solution to some problems.

Like sunlight, lasers, can pass through most gases, but unlike sunlight, lasers can cut their way through partially contaminated gases, for example, smoke. The laser will not diffuse and will maintain sufficient energy throughout the medium.

Therefore lasers can be used in environments such as:

- steam;
- smoke;
- under or through water;
- dusty environments;
- generally polluted environments.

Lasers are a monochromatic light source (single wavelength) and are polarised to travel in a single direction. This makes them ideal for travelling long distances with very little variation. They are, therefore, ideal for use in:

- measuring long workpieces – bars, tubing, girders;
- for use 'out in the field' over large areas of land;

- to measure expansion in large objects (e.g., the cockpit in Concord);
- long distance proximity sensors;
- to measure relative movements of buildings.

Vibration analysis can also be achieved by lasers. The laser is aimed so the spot focuses on either the workpiece or the tool (the workpiece in milling and the tool in turning). The spot is then reflected off the surface and is received by a TV camera. A laser pattern is received by the camera and consequently transformed into a vibration image (similar to a vibration pattern received from an accelerometer as used in tool monitoring (see Chapter One)). This vibration image can then be analysed automatically by computer and compared to a standard vibration pattern. If the two patterns deviate in any manner possibly due to a faulty tool, then an alarm can sound, alerting the operator and possibly stopping the machine.

3.8   OTHER NON-CONTACT SENSORS

3.8.1   Proximity methods (also see section 2.4)

The use of pneumatic location, whereby air switches detect displacement between the workpiece and a datum position, is a well-known method. It can indicate when the workpiece is correctly located, but the sensors can become contaminated by swarf, coolant, etc. Eddy current location devices have been developed to overcome some of the difficulties inherent in pneumatic location.

Ultrasonic, capacitive, and inductive proximity devices may form the sensing heads of calliper devices, in which the callipers are adjusted until the sensing elements are at a set distance from the workpiece. This type of size measuring device requires positioning, accurate actuation, and a further device to monitor the calliper separation distance. These systems are not favoured in industry as all the instrumentation is located within the work zone. Also, air jets are required to clean the components (particularly for the capacitive heads). Inductive heads are small and less sensitive to coolant, but can only detect metallic surfaces.

3.8.2 Microwaves

Microwaves can be directed at the workpiece, the phase angle difference between the incident and reflected waves being measured. This phase difference angle can be used to determine:

- angles of faces;
- dimensional data;
- component features.

These devices have fairly large stand-off distances; a fixed datum is required for dimensional measurement.

3.8.3 Example of a non-contact system in use in industry

Dawson (of Dudley, West Midlands, UK) manufacture piping. The range of the piping varies from cast concrete wide diameter to small, precision PVC.

The plastic pipes are extruded. Directly after extrusion the pipe enters a pressure-controlled vessel, which is filled with a temperature-controlled steam environment. This is to stop the tube collapsing and to maintain a uniform optimum shape.

It is necessary to determine the diameter of the tube as it passes through the vessel. The most effective way to achieve this is through the use of a laser. The laser scans across the diameter of the tube and accurately measures the diameter of the semi-cured tube. As it is a steam-filled environment it is very difficult to measure the diameter; but using a laser the steam makes very little difference, which has resulted in accurate control of an otherwise difficult situation.

FOUR

Miscellaneous Sensing Devices for Industrial Use

4.1 INTRODUCTION

This chapter covers commonly used industrial sensors and sensor-based systems that do not fully conform to types described in earlier chapters, but which can easily be used or adapted for use. These sensors are essential in many areas throughout a machining system and an organization. The following types of sensor are described:

(1) Pressure sensors.
(2) Inclinometers.
(3) Velocity/displacement sensors.
(4) Fluid level sensors.
(5) Sensor control units.

4.2 PRESSURE SENSORS

Pressure sensors can be used for many applications and in many environments, such as power plants, water treatment facilities, aircraft and marine hydraulic systems, nuclear testing, flight qualified systems, and a number of energy management and climate control systems.

The broad range of standard pressure products enables most pressure sensing needs to be satisfied off the shelf. There are also special configurations for particular applications.

There are many types of sensor available, each adapted for a type of application and for a certain range of pressures. These can be categorized into the following two groups.

(1) For medium-to-high pressure (ranging from a few millibars to several hundreds of bars), using a strain gauge arrangement.

(2) For very low pressures, typically measured in mm of water column. This uses an LVDT-type of reliable and economical arrangement.

4.2.1 Applications

The following are some typical applications in differing areas:

Marine

(1) For control of submersibles, such as measuring ballast pressure on each leg of an oil rig platform as it is submerged to the ocean floor.

(2) Content measuring on chemical tankers.

(3) Monitoring hydraulic oil and fuel pressures of ship-borne equipment.

Aerospace

(1) Hydraulic pressure measurement on aircraft.

(2) In-flight pressure control.

(3) Measurement of altitude (atmospheric pressure) and air speed.

Industry

(1) Process control in the food industry, dairying, pharmaceuticals, and brewing.

(2) Liquid vat or tank contents measurement.

(3) Measurement and control of hydraulic pressures on steel rolling mills.

Water
(1)   Monitor nuclear reactor core pressure.
(2)   Measure water level behind hydro-electric dams.
(3)   Power plants.

Transportation
(1)   Monitor brake and oil system pressure on locomotives.

On determining the application, there are four main types of pressure reference to consider.

Vented gauge (VG)
The measurement is referenced to atmospheric pressure. Zero is set at atmospheric pressure.

Sealed gauge (SG)
The measurement is referenced to a sealed internal reference pressure. Zero is set at atmospheric pressure.

Atmospheric pressure (A)
Measurement is referenced to an internally sealed vacuum. Zero is normally set at absolute zero.

Differential pressure (D)
Measurement is the difference between two unknown or line pressures. Measurement can be unidirectional, where one pressure is always higher than the other, or bi-directional, where the higher pressure may change from one port to the other. In differentials, consideration should also be given to the media being measured, e.g., is it dry or wet?

4.2.2   Measuring medium-to-high pressures

Standard versions of medium-to-high pressure sensors use a stainless steel diaphragm to sense pressure. The deflection of the diaphragm is transferred to a double cantilever beam by a force transfer rod. This input pressure creates strain in the beam, and this is measured by four foil strain gauges.

This type of arrangement provides a reliable and stable sensor, relatively insensitive to vibration, altitude, and shock. For specifications see Table 4.1.

Table 4.1 Specifications for medium-to-high pressure sensors

Pressure ranges	
high:	5–70 bar
medium:	0.5–35 bar
Pressure references:	SG, VG, A
Pressure limit:	x5 full pressure
Burst pressure:	x20 full pressure
Pressure media:	Liquids or gases
Temperature range:	–54° to 150°C
Acceleration response:	
high:	0.02% FRO
medium:	0.10% FRO
Insulation resistance:	500MΩ at 50V d.c.
Weight:	125–180gm

4.2.3 Measuring low and very low pressures

These pressure gauges are of the vented gauge type, absolute or wet/wet differential pressures, with a variety of fluid media.

The pressure sensing element of this sensor includes an all-welded capsule which offers low hysteresis and a constant scale factor with temperature variation. The deflection of the capsule when pressurized is measured by an LVDT displacement sensor whose core is directly coupled to the capsule. The LVDT produces an electrical output which is directly proportional to the core's motion, which is, in turn, proportional to the pressure applied to the capsule. Specifications are given in Table 4.2.

4.3 INCLINOMETERS

Inclinometers are sensitive transducers which measure level, horizontal angle, and vertical deviation, with virtually infinite resolution.

Inclinometers may also be known by other names, and other sensors may perform similar functions. Some of these are:

– electronic protractors;
– electronic clinometers.

Table 4.2 Specifications for low and very low pressure sensors

Pressure ranges:	0–7 bar
Pressure references:	VG, A, D
Pressure limit:	x5 full pressure
Burst pressure:	x20 full pressure
Pressure media:	Liquids
Temperature range:	–40° to 80°C
Humidity:	95% relative humidity
Acceleration response	
high:	0.02% FRO
medium:	0.10% FRO
Insulation resistance:	500MΩ at 50V d.c.
Weight:	300–550gm

Recommended working range

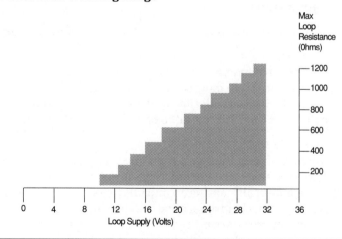

69

Modern construction of these sensors allows for their use in all weather conditions and they can withstand shock and vibration. This makes them very useful in industry, where there may be varying climatic conditions and constant exposure to shock and vibration.

The devices operate from a standard d.c. supply. Output is an analog d.c. signal directly proportional to the sine of the angle of tilt. In the level (horizontal) position, the d.c. output is zero. When tilted, the output of the sensor varies from a negative voltage in one direction of tilt to a positive voltage in the other.

The inclinometer operates as a closed-loop torque balanced servo system (Fig. 4.1). The heart of this gravity-referenced angle detector is a torsional flexure-supported moving mass system which is rugged enough to withstand severe shock and vibrations and still maintain accuracy. The servo-system electronics, torque motor, and feedback sensor are all enclosed within an environmentally sealed housing, permitting operation under hostile conditions without degrading performance.

Inclinometers are supplied according to the angle range required. With each different angle range, the inclinometer's specifications may alter slightly.

4.3.1 Applications

The following are some typical examples of an inclinometer's applications:

Woodworking environment
This provides an accurate method of establishing the correct position of equipment during installation and for in-process use, such as setting the correct angle of the table of a cutting tool.

Oil well pump control
A microprocessor-based pump-off controller uses a quantitative load and position transducer to ensure that an oil pump operates only when it is mechanically and economically feasible to do so. These sophisticated systems help determine well production capability, and control pumps operate accordingly. The inclinometer acts as the position transducer within this system.

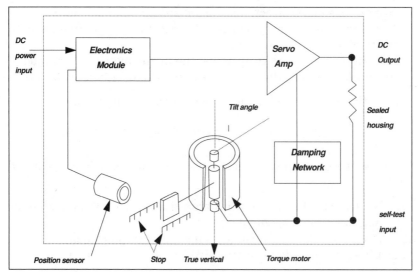

Fig. 4.1   The inclinometer

Marine
Inclinometers are used in measuring the heel angle or roll of sailboats, the pitch and roll of undersea robot vehicles, the position of television cameras, and the directional control of underwater lights. An added benefit of inclinometers is that they are not affected by water pressure.

Machine tools and heavy machinery
The level condition and angular position of machine tool tables and cutting tool tables can be accurately checked and aligned with inclinometers.

Aligning and measuring squeegee angles on screen printing presses is another good example of an application. The proper angle is important during set-up procedures, but of greater importance, once the correct angle has been determined, is the repeatability of that setting, both on the same machine and from one machine to another.

Land navigation and automobile security
Automobiles may soon rely on computers to provide detailed travel

information, such as directions and other data on road conditions and timetables. This will only be possible with the information from an inclinometer.

For automobile security systems, the inclinometer is a method of determining whether the car is jacked up, tilted, or hoisted by a tow.

Specifications for inclinometers are given in Table 4.3.

4.4 VELOCITY AND DISPLACEMENT SENSORS

There are three main types of these sensors which will be discussed in this section.

(1) Linear variable differential transducers (LVDTs) for angular and linear displacement.
(2) Magnetostrictive transducers for linear displacement.
(3) Linear velocity transducers (LVT).

4.4.1 The LVDT for angular and linear displacement sensing

The LVDT is widely used today as a measurement and control sensor, for displacements from a few micrometers to several metres. Displacements can be measured directly or where other physical quantities, such as force and pressure, can be converted to linear displacement.

Largely because they are capable of extremely accurate and repeatable measurements and operate in extreme environments, LVDTs are the optimum transducer elements for many industrial applications.

Type of measurement
Linear displacement – displacement is measured along a line coincident with the axis of the transducer.

Angular displacement – the amount of rotation about an axis up to ±60 degrees.

Table 4.3 Specifications of inclinometers

	Range		
Characteristic	*±1°*	*±14.5°*	*±90°*
Input voltage (d.c.)	±12–18	±12–18	±12–18
Input current (mA)	±25	±25	±25
Full range output (V d.c.)	–5 – +5	–5 – +5	–5 – +5
Output impedance (Ω)	15K	16K	4K
Output noise (V rms)	0.002	0.002	0.002
Non-linearity (% FRO)	0.05	0.02	0.05
Vibration rectification (s/g)	10	10	10
Thermal sensitivity (FRO/°C)	0.05	0.01	0.003

Environmental characteristics

Operating temperature range:	–18 – +71°C
Survival temperature range:	–40 – +71°C
Constant acceleration overload:	50g
Shock survival:	1.5kg
Vibration endurance	
sinusoidal:	50g peak
random:	35g rms, 20–2000Hz

Method of measurement

Mechanically linked – physical connection between the transducer and the object being measured.

Spring-loaded – projection from the core is a spring loaded against the object being measured.

Input / output

The choice between an a.c. or d.c. operation unit is determined by the application.

a.c. – for use in a.c. systems (50Hz–25KHz) or where d.c. operation is required with remote location of the signal conditioning.

d.c. – for use in d.c. systems or where portable battery operation is required; d.c. LVDTs are, in effect, a.c. units with integral signal conditioning.

4.4.1.1   Advantages of LVDTs

(1)   Frictionless measurement.
(2)   Infinite mechanical life.
(3)   Infinite resolution.
(4)   Null resolution.
(5)   Cross axis rejection.
(6)   Extreme ruggedness.
(7)   Core and coil separation.
(8)   Environmental compatibility.
(9)   Input/output isolation.

4.4.1.2   The operation of the LVDT

The LVDT is an electromechanical device which produces an electrical output proportional to the displacement of a separate movable core. It consists of a primary coil and two secondary coils symmetrically spaced on a cylindrical form. A free-moving rod-shaped magnetic core inside the coil assembly provides a path for the magnetic flux linking the coils. A cross-section of the LVDT and a plot of its operating characteristics are shown in Figs 4.2 and 4.3.

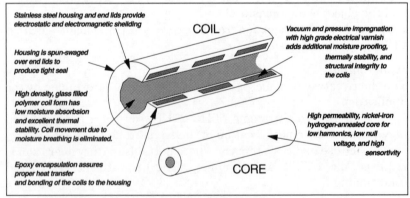

Fig. 4.2   Cut-away view of an LVDT displacement sensor

4.4.2 The magnetostrictive transducer for linear displacement

The magnetostrictive linear displacement transducer provides highly accurate absolute position measurement of displacements up to 3 metres. The sensor also measures velocity, giving feedback of both position and velocity simultaneously. Also, by adding a magnetic float, the sensor is capable of liquid level measurement.

Operation is based on accurately measuring the distance between a pre-determined point and a magnetic field produced by a moveable permanent magnet. The design and durability of the sensor make it ideal for measuring the position of moving machine parts.

Performance
This sensor is a marked improvement over potentiometers for long stroke position measurement. It possesses exceptional linearity, even

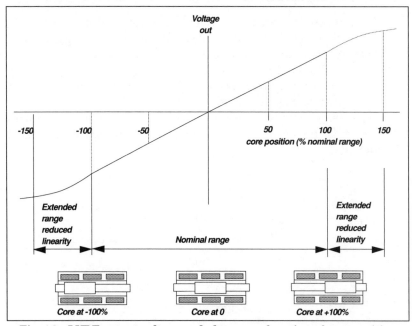

Fig. 4.3   LVDT output voltage and phase as a function of core position

over lengths up to 3 metres, with repeatablity up to 0.002 per cent of the measurement range.

With magnetostrictive technology the sensor requires no contact between parts to cause friction or premature wear. Contactless operation also provides a long life. A good stroke-to-length ratio provides a more compact long-stroke position sensor.

Selectable input / outputs

This displacement sensor is d.c. powered by ±15Vd.c. or 24Vd.c. By providing an absolute output, no data are lost should the power be interrupted. The long-stroke sensor offers a universal analog position output, as well as a velocity output. Therefore, one unit can be used for multiple applications. A digital pulse variation of the sensor is also available for users requiring a gate pulse output.

4.4.2.1 The magnetostrictive displacement sensors operation

The electronics in the sensor periodically send short-duration current pulses ('start' pulses) down the wave guide. When the magnetic field accompanying the current pulses intersects the magnetic field of the moveable ring, magnetostriction causes a momentary torsional strain in the wire. This strain travels down the wire and is converted back to an electrical pulse ('stop' pulse) in the head of the sensor.

By measuring the time difference between the start and stop pulses (see Fig. 4.4), the position of the magnetic ring can be precisely determined.

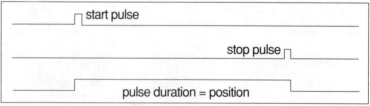

Fig. 4.4

4.4.3 The linear velocity transducer (LVT)

The linear velocity transducer (LVT) produces an electrical output directly proportional to the time rate of change of rectilinear displacement (linear velocity). It is a rugged self-contained device which provides a simple and accurate method of measuring the linear velocity of an object as it is displaced from one position to another.

An LVT consists of an enclosed coil assembly and a separable permanent magnet core free to move axially within the coil assembly without mechanical contact. LVTs are generally used for dynamic measurement on mechanical devices or structures, especially where direct determination of linear velocity is desired. Usually the low mass magnetic core is attached to the moving member, while the coil assembly is attached to the fixed member.

An LVT can be mechanically coupled to a suitable LVDT to give independent position and velocity information for stability feedback in servo systems. Other typical applications include structural vibration measurements, seismic instruments, actuator speed controls, and dynamic balancing equipment.

4.4.3.1 The operation of the LVT

A typical linear velocity transducer, shown in Fig. 4.5, utilizes a small cylindrical permanent magnetic core of high coercive force, enclosed in a non-magnetic sleeve. This sleeve has internally threaded inserts at each end to facilitate attaching the magnetic core to a test member. Two identical coils are wound side by side on a coil form that is coaxial with the magnetic core.

By Faraday's law of induction, a voltage proportional to the rate of change of magnetic flux is developed in each coil as one pole (end) of the magnet moves axially inwards. Therefore, from this it is possible to determine the velocity by the voltage produced.

4.4. Applications for velocity/displacement sensors

Injection moulding
Based on its high resolution and repeatability, the displacement

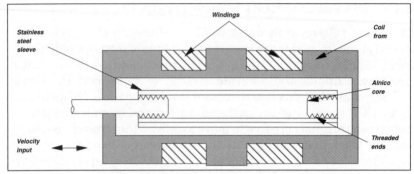

Fig. 4.5   Simplified LVT cross-section

sensors replace potentiometers for platen positioning and monitoring in plastic injection moulding equipment and die casting machinery. The non-contact sensor offers long life, even in hostile environments.

Ovens

With a pressure sealed probe capable of withstanding temperatures to 85°C, these sensors are used to operate oven doors to specified opening height and closure requirements. In addition to providing a very repeatable and reliable measurement, the sensor may provide a 4–20 mA output to interface with a remote PLC.

Hydraulic cylinders

These sensors are capable of performing in hydraulic fluid, providing load, position, and velocity feedback in the cylinder. The sensors can withstand upto 5000 lb/in² and, therefore, can be packaged within the cylinder as a closed-loop feedback device. The sensor maintains the cylinder position at power-up, eliminating the need to initialize the system.

Machine tools

Using a float, these sensors can measure coolant level in the sump of machine tools. Additional information can be acquired, such as the tool position and feedrate, and the position of the workpiece.

Grinding machines

The sensors provide, electrically, an indication of the grinding wheel position. A controller can adjust the speed of the grinding wheel to compensate for any change in diameter. Measuring to a non-linearity of 0.05 per cent of full scale, the sensor provides the highly accurate data necessary for this application.

Materials handling
In the control of automated multi-shelved materials handling systems, the sensors offer strokes of up to 3 metres and repeatability of 0.002 per cent, providing an ideal means of accurately positioning stored parts.

4.5   FLUID LEVEL SENSORS

The measurement of level is critical to many manufacturing operations; for example maintenance of the correct oil lubrication level, or the correct volume of coolant.

Due to the wide range of liquids (and solids) which need to be measured and monitored, their varying properties, and the storage conditions within the vessel, broad specifications are required for level sensing instrumentation.

One of the most commonly used level sensors is a float encased in a rigid housing (see Fig. 4.6).

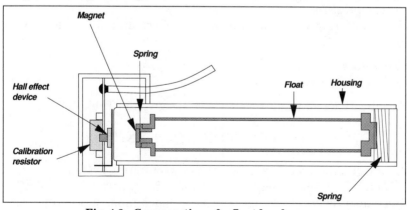

Fig. 4.6   Cross-section of a float level sensor

4.5.1 The operation of the fluid level sensor

In the heart of the sensor is a suspended float element whose buoyancy causes slight relative motion. This small movement of the magnet-topped float, in relationship to a Hall Effect device, provides a positive indication of the fluid level.

The float is suspended within the mounting tube by a pair of springs, top and bottom. The springs, which are radially rigid, but relatively flexible longitudinally, position the float and allow smooth movement without the usual stiction, hysteresis, and wear. The fluid sensor can be adapted to a wide range of measuring requirements by changing the spring constants and lengthening the float and mounting tube.

4.5.2 Applications of the fluid level sensor

Truck and automotive engines
For oil, fuel, and coolant volumes/levels.

Transmissions
For oil levels

Hydraulic reservoirs
For hydraulic fluid levels.

Air compressors
For monitoring the volume of water collected, and for lubrication levels.

Generators
For fuel, lubrication and coolant levels.

Specifications for fluid level sensors are given in Table 4.4.

4.6 SENSOR CONTROL UNITS

A control unit is the 'brain' of a sensor monitoring system.

It is an electronic device which automatically performs monitoring and supervision of a wide range of machining, processing, and

Table 4.4   Specifications for fluid level sensors

Repeatability:	±3mm
Output:	Radiometric analog
Power:	5Vd.c.
Operating temperature:	−100 to 165°C
Case:	Liquid crystal polymer
Size:	From 30mm x 10mm diameter to 2000mm x 200mm

handling operations.

The controller handles analogue signals sent from sensors detecting changes in force, torque, pressure, power, and the like. It can continuously collect and measure data from many different operations (usually upto 1000 maximum) from an analogue sensing device. It can analyse, determine, and take action, depending on the specific criteria related to the operation.

The controller will determine any abnormal process condition in a wide range of applications (see Fig. 4.7).

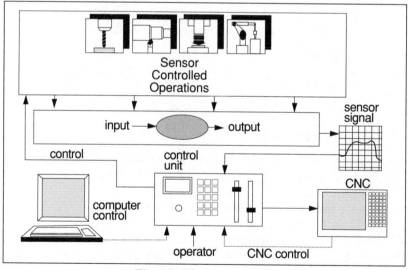

Fig. 4.7   The control unit

4.6.1 Applications

In metal cutting operations the controller detects collision, tool breakage, and wear. This reduces subsequent damage to tooling, workpiece, and machine. It can also determine if tools or workpieces are missing or incorrectly clamped.

In assembly, robot handling, and pressing operations, force can be constantly monitored to detect collisions and to ensure product quality and security.

High levels of automation are offered with a controller. Operators can run machines without constant supervision at optimum production rates with the assurance that the process is continually under precision control.

Controllers are highly suited to production environments where secure work flow, protection of machines, and the reduction of manual supervision are demanded.

Table 4.5   Specifications for sensor control units

Nominal input sensitivity:	0.1 mV/V
Maximum inputs:	up to 30 sensors
Maximum number of operations per channel:	1000
Response time:	3 ms
Sampling rate:	4000/s
Power requirements:	115 to 250Va.c. 120VA
Serial interface port:	RS 232
Protection class:	IP 55

4.6.2   Typical features

(1)   Modular design, enabling many channels for monitoring.
(2)   Processing capability to handle up to 1000 different signals per channel.
(3)   A wide range of application-related 'learn' modes.
(4)   Self-test functions.
(5)   Built-in service functions.
(6)   Reliable monitoring of large force variations within a single operation.

(7) Bar graph display for each channel.

Specifications for sensor control units are given in Table 4.5

FIVE

Condition Monitoring and Predictive Maintenance

This chapter describes the predictive maintenance techniques currently available, and their application in industry. Reference is made to a variety of industrial situations, indicating the importance of condition monitoring and predictive maintenance methods in maintenance engineering management.

5.1 INTRODUCTION

Good condition monitoring of machines and machine systems is essentially the collection of reliable, repeatable data. These data are then used to produce maintenance information to achieve optimised machine performance. Clearly it is the aim of all maintenance engineers to remove all unscheduled, unforeseen break-downs, raising the production time and hence the profit.

Predictive maintenance or condition monitoring (as it is more commonly known) was first born in the American Nuclear Industry in the 1960s. More recently the Department of Trade and Industry in the UK has produced documents which aim to tackle the problem of maintenance efficiency, thus recognizing the importance of predictive techniques. Following this lead the Institute of Mechanical Engineers is now looking at the subject of maintenance in the manufacturing and process fields, with a view to producing guidelines in this field.

5.2   PREDICTIVE MAINTENANCE AND SENSORS

Predictive maintenance requires the monitoring of a machine or a machining system. This is achieved through the utilization of sensing devices. Sensors are integrated within a machine or machining system to give the required information on the present condition of the system. Many different types of sensors may be required to do this, such as force sensors, pressure sensors, vibration analysers and so on.

The most common and well-used sensors in use in industry for these purposes are described and analysed in the preceding chapters.

5.3   MAINTENANCE METHODS

Currently there are three areas available to maintenance engineers.

Breakdown maintenance
This was, until recently, the most commonly applied approach to maintenance, with the major disadvantage being that it could not be predicted. If a system breaks down frequently, this indicates that machinery has failed to be successfully maintained during the normal running operations, and has reached a point at which it is incapable of further action without maintenance. This 'run to fail' maintenance, therefore, incurs 'panic management', as a large loss of production is usually incurred. A further problem of this approach is that secondary damage may be caused by the initial failure.

Planned maintenance
Maintenance is planned to occur at regular intervals of time. This generally is successful in eliminating most long-term failures (large breakdowns), but it has the disadvantage that it usually proves quite costly, for two reasons.

(1)   Loss of production during the machine downtime.

(2)   Components may be removed, replaced when in fact they were quite able to remain in service.

Clearly, maintenance planned in this way makes poor use of the staff available.

Predictive maintenance
This has several advantages over the other two methods of maintenance. Machinery can be assessed for maintenance while it is still running. The assessment uses non-invasive techniques which are accurate and relatively inexpensive to install. From the running condition assessment (diagnostic) maintenance actions can be planned to take place at a time convenient to both machine and maintenance staff. This method enables informed decision making and utilizes staff to its full potential.

5.3.1   The direction of maintenance

No one method will suit all processes. What is necessary is, to a greater or lesser degree, a combination of all three approaches, with a full knowledge of all the possibilities that are available. Figure 5.1 shows the effectiveness of a maintenance programme.

In the majority of cases the maintenance of manufacturing will need to be predictive-driven, with elements of planned maintenance included. For example, boiler and pressure vessel examinations need to be done on a time-based system, owing to safety precautions – and the law!

Therefore, companies must adopt different philosophies on maintenance. Many large, continuous-process industries, such as the oil, gas, chemical, and paper industries, adopt a predictive maintenance philosophy, but backed up by time-based examinations. It must not be forgotten that all predictive maintenance examinations need to be carried out regularly on a planned schedule – it is only at the maintenance action stage that there is significant change.

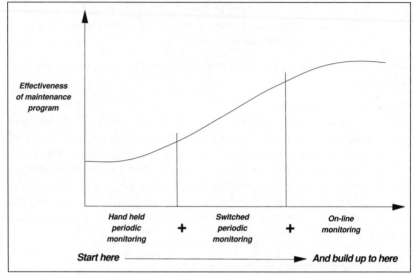

Fig. 5.1   Maintenance programme effectiveness

5.4   THE DECISION TO MONITOR

Accepting the decision to monitor gives rise to further decisions.

(1)   Initially, which equipment to monitor.
(2)   At what intervals to monitor.

There are a number of standard methods by which the criticality of a machine to the smooth running of the plant can be assessed.

5.4.1   Criticality

If any one machine on a single line process breaks down and causes the product not to be made, the effect is catastrophic. Each machine on a single line process is critical to the final product. Therefore, all machinery on a single line process shares the same criticality unless any of the following features apply.

Increased criticality
(1) High capital cost plant usually means expensive spares.
(2) Foreign plant may require parts from overseas, which may be difficult to obtain quickly and may be expensive.
(3) If a plant is old and has a higher-than-average risk of problems, its criticality and its need to be monitored are increased.
(4) Machines operating at 100% per cent design output need a high criticality rating as they are exceeding their design capabilities and creating an excess stress on the machine.
(5) If there is a safety consequence of the machine failing, then the machine's criticality must be increased, even if it is not part of a single line process. For example, ventilation fans in mines.
(6) Machines which are inaccessible and awkward to regularly maintain have an increased criticality rating.

Reduced criticality
(1) Availability of stand-bys reduce the machines criticality considerably.
(2) Monitoring capability.

Using a combination of all the factors above, a scoring system is arranged so that the items of the plant which are most critical score the highest points. A list can thus be drawn up for any plant, which identifies the most critical items for monitoring. Once this has been done then the monitoring techniques to be used can be selected.

5.5 COMMON MONITORING TECHNIQUES

5.5.1 Vibration analysis

Probably the most common and universally acceptable technique for condition monitoring for rotating machines is vibration monitoring and analysis. To achieve successful repeatable condition monitoring it is essential to collect good data from the machines concerned. This is achieved by:

- selecting the correct point to monitor from;

   – fitting the most compatible vibration sensing device.

One common sensing device which is commonly used is an accelerometer stud. This is a retro-fitted device, which can also be hand held, which determines the vibration by utilizing an accelerometer. Figure 5.3 shows the hardware in vibration analysis, using accelerometer devices.

The units for collecting vibration data can, on most data collection systems, be selected by the user at a global level, so no changes have to be made at the machine level. The most commonly used units are velocity units, for general readings. If high frequencies are required to be monitored, then acceleration units are used. The sensor used for the collection of vibration data will usually be an accelerometer, collecting a signal in changing volts proportional to the acceleration signal seen, giving, typically, a sensitivity in the range of 10–100mV/ kg^{-3}. The collected signal is a time waveform of mV vs time.

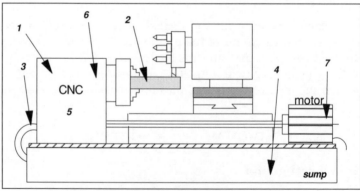

Fig. 5.2 Preventative maintenance techniques
    (1)   **Vibration analysis**
    (2)   **Temperature monitoring**
    (3)   **Flow and pressure analysis**
    (4)   **Oil analysis**
    (5)   **Acoustic emission**
    (6)   **Thermography**
    (7)   **Motor monitoring**

5.5.1.1 Data collectors

There are many data collectors available in the market place, with new versions becoming available all the time.

The collector will take the raw signal from the sensor (accelerometer) and carry out on it a number of operations.

For predictive maintenance the data are usually presented in the form of a spectrum or trend. A spectrum is obtained from the raw sensor data by integrating and carrying out a fast Fourier transformation (FFT). This produces data which is frequency-based rather than time-based, and tells the trained eye exactly what a particular machine is doing at that time. The amplitude levels in the spectrum represent discrete machine characteristics over the range of frequencies sampled. Figure 5.4 shows a sample spectrum.

5.5.1.2 Special interpretation

For most processes and manufacturing plants, the size of the spectrum for a sample is simply chosen by taking in at least half of the

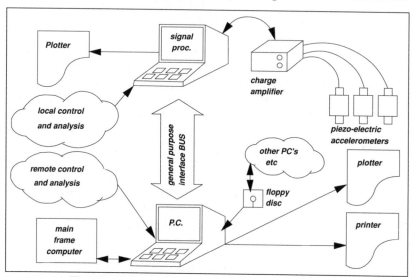

Fig. 5.3   Information system for vibration analysis

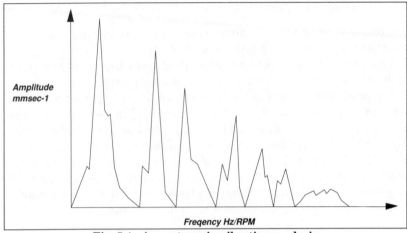

Fig. 5.4 A spectrum in vibration analysis

running speed up to around 30 times the running speed. This will include more than 90 per cent of all bearing failure frequencies, and sufficient harmonic information to be reasonably accurate for predictive trends.

For example, if a motor was running at 1450 r/min, we would take the minimum frequency to be 5Hz and the maximum at 750Hz. Therefore, a spectrum containing the frequency range 5–750 Hz would be taken. Within this, a number of lines, typically 400, but up to 1600, would be taken. This gives improved resolution for the solution of complex, closely matched components. It is usual to take a high band reading which will be in acceleration units, with frequencies from around 1000 Hz to 15 000Hz.

The spectrum forms the heart of the effective condition monitoring system. Real maintenance information is acquired from spectral information. By analysing the amplitudes of discrete frequencies a number of problems can be identified and charted or monitored. For example:

– a machine which is going out of balance will show an increase in its running speed peak;

- an alignment problem in a coupling will show itself as a changing amplitude at two times running speed;
- a problem with blades on a fan will show as a changing amplitude at the running speed, times the number of blades;
- for bearings, comprehensive tables are available which calculate the bearing failure frequency for all common applications.

5.5.1.3 Trend plot

For any discrete frequency a trend plot can be produced to show how the frequency changes as a function of time. Any fault which occurs in a machine can thus be followed, giving time to plan maintenance actions to prevent any damage to the machine and to give the minimum interruption to production.

5.5.2 Temperature monitoring

By taking regular readings of temperature of the machine's components it is possible, in the same way as with vibration, to plot a deterioration through the rise in temperature. Most bearings will give a rise in temperature as they begin to fail. Temperature monitoring has, for many years, been used to protect large white metal and plain bearings, although the time-to-failure on these types of bearings, from the point where a temperature change is sensed, can be very small. This is where an alarm system is really required, to give an immediate warning of the problem.

For example, mine-winders and fans are most often equipped with this type of bearing. A temperature rise on some oil-filled rollers might also indicate a reduction or even an absence of lubrication.

Where both temperature and vibration are monitored at the same time it is typical to see the vibration trend change slightly in advance of the temperature, indicating the vibration sensing is actually picking up the mechanical process within the machine components which subsequently gives rise to the temperature change – a pitted race in a bearing, for example.

5.5.3   Flow and pressure analysis of hydraulic and coolant circuits

To carry out condition monitoring on some hydraulic and cooling equipment it is necessary to monitor flow and pressure performance. Both these factors need carefully positioned transducers/sensors from which all the flow and pressure readings can be taken over a period of time. A flow reduction indicates that the system's health is failing – for example, worn impellors on pumps, or leaking or burst seals.

Therefore by taking flow and pressure performance figures and trends, it is possible to check in a non-invasive manner the condition of the internal components of the circuits.

Cooling water pressure change is important for any system operating with a heat exchanger. By monitoring the pressure and temperature changes it is possible to know the rate of fouling and also the thermodynamic efficiency of the exchanger.

5.5.4   Oil analysis

Oil analysis or debris monitoring is one of the most popular and simple methods of introducing condition monitoring into industry. The primary objective is to detect early signs of excessive wear and imminent failure by picking up signs of debris within the oil. An early warning can thus be gained of incipient failure. There are, however, a number of drawbacks to this type of monitoring. It is not possible, at the moment, to carry out the technique on grease lubricated bearings or on circulating oil systems (e.g., steel mill gearboxes). Where oil is passed over a number of gears and bearings it is possible to detect signs of steel debris, but not to pin-point the exact location of the debris source. As with temperature monitoring, oil analysis provides a very good complementary measure to vibration monitoring, and the source of the debris should be detected by vibration monitoring.

Examples are in food processing where oil analysis is used to detect food particles in oil, and coal cutting machines, which are particularly susceptible to dust and water in the oil, and to seal problems.

Ten oil test options

Oil analysis will normally involve one or more of the following tests.

(1) Kinametic viscosity at 40°C and 100°C: picks up contamination by other lubricants, petrol, diesel, or solvents, and severe oxidation.

(2) Water content: may indicate cylinder head gasket failure or seal failure, for example. Also checks the condition of special water-containing hydraulic fluids.

(3) Insolubles: used to ascertain filter efficiency and seal effectiveness – it may also indicate insufficient fuel combustion (fuel soots in crankcases).

(4) Total acid number: indicates prolonged high temperature use or oxidation of a lubricant.

(5) Total base number: used to test alkalinity (normally in crankcase, where built-in reserves of alkalinity are depleted by the weak acid by-products of some combustion processes).

(6) Flashpoint: the presence of fuel in the lubricant can indicate fuel injection system faults or intermittent leakages.

(7) Particle quantifier: picks up potential failures caused by non-abrasive wear (fatigue pitting, scuffing, and spalling) which tend to produce ferromagnetic particles, larger than those produced by wear.

(8) Optical microscopy: examines particle shape, size, and colour, to get an idea of the source and the nature of the wear (fatigue, scuffing, etc.).

(9) Elementary analysis by inductively coupled plasma: detects, identifies, and quantifies wear elements in the form of parts per million. It can also check that the levels of additives is correct (e.g., antiwear additives).

(10) Particle counting: where hydraulic and turbine systems require super-clean lubricants, the cleanliness of the system is expressed in the form of a particle count. The result reflects on the efficiency of the seals and filters.

5.5.5 Acoustic emission

Acoustic emission is a well-used technique for pressure vessels which is gaining popularity in other industrial areas.

There is a narrow field of application for acoustic emission techniques, but it has been used to monitor bearings, especially for slow speeds. For example, monitoring the bearing speed in the refractory and cement manufacturing industries, monitoring dock levelling bearings at ports, and monitoring most slow speed loaded bearings.

Acoustic emission looks for very high frequency noise pulses (in the 0.1–1 MHz range). It is the changing of the amplitude and the number of these pulses which determines the health of the bearings.

5.5.6 Thermography

Thermography has been used to measure hot spots in electric panels, without removing the covers or disconnecting the power. This technique has, over the years, been improved and can now be used to check bearing health. However, the technique has the same 'after the event' problems that are found with temperature monitoring.

An example of its use is with very high voltage cable joint examination to determine the quality of the join.

5.5.7 Motor monitoring (see example 1 below)

Most industrial manufacturing processes will be driven by an electrical prime mover. In many cases this will be an induction motor. There are two techniques which can be applied to induction motors to give an early indication of any internal problems. The bearings in the motor can be monitored by using vibration analysis with an accelerometer. For the stator and the rotor winding to be monitored, it is necessary to have a current supply tapping, or a search coil on the motor. From the current supply taken on one phase of the supply, by using a current/power sensor or a Hall effect probe, information can be gained about the rotor condition, static and dynamic eccentricity, loose rotor bars, and shortened turns in a wound stator. This information is acquired by analysing the amplitude of the side bands on the current supply spectrum at the slip frequency. Each particular problem has its own characteristic spectrum.

By fitting search coils on the ends of the motor it is also possible to monitor the changing axial flux induced in the coils. By analysing the spectrum of the signal a number of faults can be detected.

(1) Broken rotor bars.
(2) Starter winding interturn short circuits.
(3) Wound rotor short circuits.
(4) Interturn short circuits on un-energized windings of double wound machines.
(5) Loss of phase.
(6) Negative phase sequence in supply lines.
(7) Eccentric running.

Motor monitoring by either technique gives a very good indication of the problems in the motor without taking it out of service. For example, in a large water pumping station, axial flux can be measured to detect rotor and stator faults.

5.5.8 Examples of condition monitoring methods

Example 1   Condition monitoring of a machine tool axis drive via the motor

Abstract
In this case a built-in axis servo motor current sensor is used to monitor the health of the axis drive of a vertical milling machine. The axis drive consists of a d.c. servo motor, ballscrew, slideways, and a transmission belt (see Fig 5.5).

A health index (HI) of the axis drive system is established and used as the monitored parameter. It measures and checks out the health of the axis drive (both the servo and mechanical systems) under a no-load condition. A fault dictionary is set up to facilitate the application of a pattern recognition method of fault diagnosis, such as the cross-product and the nearest neighbour rule methods, to rank the most probable faults.

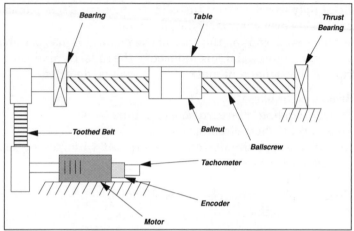

Fig. 5.5   An axis mechanical drive system

This example is adapted from a study at Cardiff University by S. M. Hoh and J. M. Williams.

Introduction

Condition monitoring is an essential requirement for any advanced manufacturing system. A machine tool is a fundamental element of such a system. At the heart of any machine tool are the axis drive systems. The condition of these relate to the condition the working machine.

The condition of an axis drive system is monitored by measuring the steady state and transient behaviour of its d.c. drive motor current. By using particular characteristics of this behaviour 'health indexes' can be established to provide an indication of the condition of the system. These indexes also provide a means of monitoring drift in the health of the system, enabling efficient condition-based maintenance to be implemented.

Axis drive unit

The machine tool used here is a Devleig/Wadkin V4-6 flexible machining cell. The *x* axis drive system consists of the following major assemblies:

- a servo unit;
- a ballscrew;
- a slideway (table support).

The servo unit includes two adjustable resistors controlling gain and balance, which are set to the customer's needs. Drift in these resistor values affects the performance of the machine, particularly in contouring operations. The other assemblies are mechanical in nature and are affected by wear and tear and the ingress of foreign matter.

Data collection and processing
There are three signals readily available from the drive systems. These are the command signal from the command controller, the tachometer feedback, and the motor current. Of these, the motor current is the most sensitive to parametric changes in the four assemblies. When the table is commanded to move, the motor current exhibits both transient and steady state features. It was found that the transient response was useful for monitoring the gain and the balance of the servo, whilst the steady state response was sensitive to the mechanical parameters of the drive. The sampling rates required differ for both states. All samples were taken with the machine table unloaded.

Health index
Under healthy conditions, data were collected over 30 runs. The maximum and minimum values of the 30 runs at each sample point were noted. At each sample point the average and maximum values were used to obtain an overall average response together with the associated maximum envelope; then a health value was calculated.

Condition monitoring and fault diagnosis
As the health index values are periodically calculated they will form a time series indicating the health record of the system. This feature readily enables a condition monitoring scheme to be implemented where maintenance action can be triggered at a predetermined value of the health index.

When maintenance action is triggered the latest dynamic response can be used in a pattern recognition sense to diagnose the most likely cause of the loss of performance.

Example 2   Ballscrew wear monitoring

Introduction

Ballscrews are extremely efficient rotary-to-linear motion converters, which are extensively used in machine tool drive systems. A useful characteristic of the ballscrew system is that it can be pre-loaded. Pre-loading of the ballscrew eliminates backlash and increases its stiffness. Wear in the ballscrew system will reduce performance by gradually degrading the pre-load, initially to a point where the unit's stiffness falls outside its specification, and ultimately, when the pre-load is entirely lost, to the point where backlash occurs. This characteristic makes pre-load an ideal parameter for assessing screw condition.

The sensors used to measure the variation in pre-load force in the ballscrew consist of a semi-conductor strain gauge bridge circuit. Four

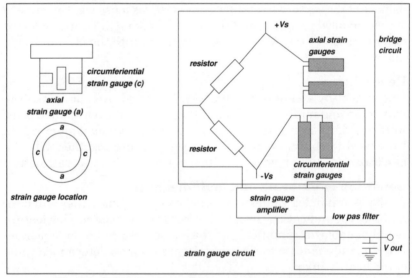

Fig. 5.6   Ballscrew transducer arrangement and location

strain gauges are evenly distributed around the external pressure faces of the ballscrew nut. This arrangement is shown in Fig. 5.6.

The strain gauges are mounted axially and circumferentially on the ballscrew nut. They measure the strains induced in the nut, either when it is compressed by the pre-load, or by any external force acting on it. An example of an external force is the slideway friction force.

A pre-load signature can be obtained by calculating the mean output of the sensor produced as the ballscrew nut travels along the screw. A slideway signature can also be obtained from the difference in the output of the sensor.

To demonstrate that the strain gauge arrangement used can detect a change in the slideway condition, a simulated change in slideway friction force was made. It was seen that the signature of the slideway changed, whereas the signature of the ballscrew remained unchanged.

The monitoring procedure involves driving the ballscrew nut from one end of the ballscrew to the other in both directions. The data collected over a period of time could then be analysed and used to determine the trend of the health of the ballscrew and the slideways. This method is simple, cheap, effective and can be retrofitted.

This example is adapted from a study at Cardiff University by P. Thorpe and K. Mart.

5.6   ON-LINE OR OFF-LINE MONITORING?

Having carried out a study to ascertain those items which will best benefit from condition monitoring, it now remains to choose the type of system to use to monitor the machines. From the range of standard methods and products, a combination is usually required to fit required monitoring to a particular situation. There are, though, two global descriptions of methods of monitoring: these are on-line and off-line monitoring.

5.6.1   On-line monitoring

On-line monitoring is the acquisition of maintenance data using a computer-based system, revolving around real time data acquisition and processing, giving almost immediate warning if any of the monitored parameters fall outside pre-configured levels. The control computer for the system usually resides alongside any plant process control, but provides much different data, although most systems can be configured to acquire and log process data for more refined analysis alongside vibration and temperature data collection.

5.6.2 Off-line monitoring

Off-line monitoring is the collection of data using a portable data collector. The most popular parameter to measure, and that which most often gives the most immediate results, is vibration data. Data are collected by using one sensor and moving the sensor and data collector from point to point. This is obviously different from on-line monitoring, where the sensor must stay in place to provide instantaneous data acquisition.

Both off-line and on-line systems require a degree of manipulation skill in setting up the systems, but once this task has been done they can be run by qualified technicians, who will refer any borderline situation to others with greater skill in the analysis and interpretation of the data.

Acquiring the level of skill to input data information into most packages of software is largely gained by studying large volumes of vibration data, and applying this knowledge to set up the new database.

5.6.3 Advantages and disadvantages of on- and off-line monitoring

The advantages and disadvantages of on- and off-line monitoring are outlined in Table 5.1.

On-line condition monitoring does not provide a core for maintenance, but it helps to focus maintenance attention immediately on any impending problem areas accurately, predicting machine failures.

5.7   DATA COLLECTION

Whatever sensor or system is used to acquire data, then the information must be collected and analysed in an intelligent way to determine what is actually happening.

This information can be collected and used in many different ways. One way, which is very common, is the use of SPC. This stands for Statistical Process Control. It is a tool used to help understand and control processes.

Control, in this context, is assessed through the use of charts. These charts are based on the normal curve (see Fig. 5.7). They have lines (known as control limits) that match the 99.7 per cent lines on the curve. If the process goes beyond these lines, then it must have changed due to some special cause. 'Process' is used, in this context, to describe the combination of the equipment/machine being used, the people involved, the methods/procedures being used, the materials being worked, and the physical environment. The use of statistics for control in SPC is based on a few simple theories:

- that all things have variation;
- that this variation is a natural part of all processes in normal circumstances;
- that this variation can be quantified (in the shape of the normal curve – Fig. 5.7);

Table 5.1   Comparison between on- and off-line monitoring

On-line	Off-line
expensive	cheap
no duplication	duplication
no redundancy	redundancy
single line process	many lines
many maintenance staff	no maintenance staff
maintenance free	intensive maintenance
screen for further work	more work all of the time
installing for retrofit is difficult	retrofit easy

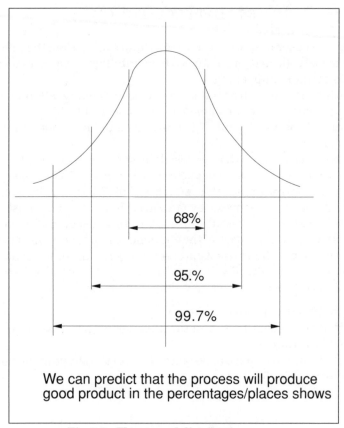

We can predict that the process will produce
good product in the percentages/places shows

Fig. 5.7   The normal distribution curve

- when quantified, predictions can be made as to what is likely to
  happen in the future (probability), this being the percentage
  under the curve in Fig. 5.7;
- when something changes in the process, the change will not fit
  with the predictions, therefore, the change will be spotted (this
  change is known as a special cause and shows that a problem has
  occurred).

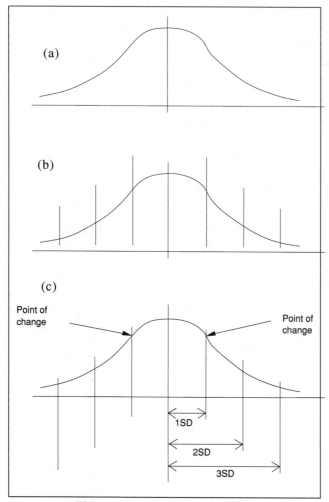

Fig. 5.8   Standard deviation (SD)

5.7.1   The normal distribution curve and standard deviation

The normal distribution has a central line which represents the average of the data that forms the curve. This average is called the mean, or the arithmetic mean (see Fig. 5.8(a)).

Lines either side if the mean are of equal spacing; this space is called the standard deviation (SD). The SD can be easily calculated using a formula given below (see Fig. 5.8(b)).

The width of one SD can be expressed as the part where the slope of the curve changes from increasing to decreasing, and vice-versa. The curve can now be divided in SDs (see Fig. 5.8(c)).

5.7.2 Control limits

As the data produce a normal distribution, it is necessary to use them in a constructive way, so that conclusions on the parameter being measured can be shown.

The data may be used by utilizing control charts, the key tool of SPC, and the normal distribution is the link between the raw data and the control chart.

Looking at Fig. 5.9, which shows a normal distribution curve, it is possible to turn this curve on its side, giving the start of a control chart. By projecting out the 'central line' and the two lines which represent the + and – 3 SD, it is possible to use these lines as control lines. (Predictions can be made of what will happen under the normal curve, and these predictions carry over to the control chart – for

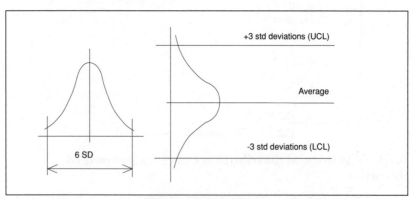

Fig. 5.9 Control limits

example, using these control lines it is known that 99.7 per cent of data will fall between the control limits if nothing changes, so if some data falls outside these control limits then it is likely that something has changed and it may be necessary to take action to correct such a change.)

To apply this to a manufacturing process:

- study the process by taking sample data over a period of time;
- use statistics to find out what the normal distribution curve is like;
- turn the curve on its side to make a control chart.

5.7.3 Control charts

Collected data will be expected to fall within the 'control limits' 99.7 per cent of the time. The same distribution predicts that 68 per cent (or two thirds) of data will fall in the middle area around the central line.

Figure 5.10 shows data from a process in the form of a control chart, showing data which are OK (in control) and data which show that

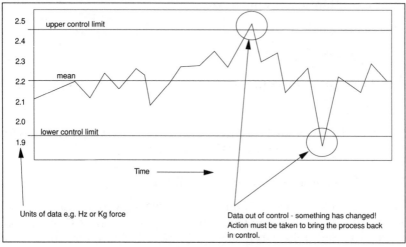

Fig. 5.10   The control chart

something has changed (out of control) for example, the gears have worn in a gear box, producing excessive vibration picked up by an accelerometer.

Using SPC, data from a sensor or a system can thus be used to monitor and control a process.

5.7.4 Data collections equations

Average mean

$$\overline{X} = \frac{X_1 + X_2 + \ldots + X_n}{n}$$

Standard deviation

$$SD = sum(X_i - X)^2/(n-1)^{1/2}$$

Control Limits

upper control limit: $UCL = \overline{X} + A_2\overline{R}$

lower control limit: $UCL = \overline{X} - A_2\overline{R}$

(A_2 is a constant for a sample size of 5)

Average range

$$\overline{R} = \frac{R_1 + R_2 + \ldots R_n}{n}$$

R is the range of the data sample.

5.8   CONDITION MONITORING/PREDICTIVE MAINTENANCE AS USED IN INDUSTRY

5.8.1   Paper manufacturers (Challenge)

The paper industry is a 24 hour process with very high capital costs. Most mills run for a number of weeks between short (6–8 hours) stoppages, and quite often these are missed. Therefore, there is a great requirement for knowing exactly the condition of the machines at all times. In the past, using a portable data collector based system enough warning was generated to allow the mill to run to a maximum before taking maintenance action. More recently, monitoring has been carried out using a continuous on-line system to give immediate warning of impending failures, allowing much more accurate maintenance planning.

At the mill producing fine writing paper a final screen was selected for monitoring. The screen was chosen because of its high criticality rating. It was a single component with no stand-by; the machine is also of foreign manufacture and spares are on long lead times and expensive. Vibration monitoring was chosen initially in the off-line mode to find any immediate problems, with a view to permanent on-line monitoring in the future. The results of an initial vibration survey showed special evidence of some degree of alignment problem, indicating internal damage to the screening elements. On examination of the screen it was found that there were broken screening elements which were impairing the screening efficiency and ultimately would have led to failure. A regular monitoring programme for this item has now been implemented.

5.8.2 Coke works (BCF)

This company produces coke oven fuel. Gas is drawn from the top of the ovens using steam turbine driven gas exhausters. If any problems occur to these machines there is an immediate health and safety hazard to the operators on the ovens, and an immediate loss of product as the coke may be unfinished and, therefore, unredeemable. These machines have, therefore, a high criticality rating and a predictive maintenance programme to replace a costly routine annual stripdown and repair was instigated.

A programme of vibration and bearing temperature monitoring was instigated. An initial survey detected a level of imbalance in one machine, which was then carefully monitored for deterioration until the next planned service. Another machine had bearing problems which led to an immediate shut-down and the replacement of the bearings; a complete collapse of the bearing was thus prevented, and costly secondary damage was avoided.

SIX

Discussion

6.1 SENSORS

Tool wear and failure is relatively well served for in-process measurement. There is still some controversy about the sensitivity of some of the different techniques, but machine tool manufactures claim that there are reliable sensors currently available. These are in the power, torque and force measurement areas, but more attention should be given to sensing nearer to the action by using the most appropriate sensing devices, with special attention given to optical imagery and vibration.

The methods most commonly used by manufactures are tool forces, measured at the bearing housing. This is a reasonably easy measurement to make, and the force sensed, especially where more than one component is measured, has sufficient sensitivity to indicate tool wear. The measurement of force has long been of acceptable accuracy, but the use of force tool dynamometers at the tool itself is not popular with manufacturers.

The other commercial methods of power and current measurement are, without doubt, attractive because of the convenience and cheapness of the measurement. These techniques receive the most support and are generally acceptable in most cases. However, it must be realised that these sensors are further from the 'action' than, for example, force sensors.

Most indirect methods have a 'learning curve', where the system has to learn what to expect, and what is normal and abnormal. For example, in video camera analysis, the image analyser must distinguish between different types of parts, and this may be achieved through learning the geometric features and colour.

There is great scope for improving the application of pattern recognition or signature analysis principles to the data emanating from sensors. The use of computers in analysing the data quickly enough to predict failure is also very important, and with the development of microelectronic systems, this will begin to play a much larger role in these fields.

Many problems in the area of alignment, position, and location have been solved with the advent of the touch trigger probe. This has provided the machine tool user with a means of accurately positioning and aligning components and tools. The controversy remains about its reliance upon the machine tool slides, but this difficulty has been eliminated by the use of laser devices, since these provide absolute measurement. However, these laser systems need to be developed further and reduced in price for direct fitting to machine tools.

The feasibility of measuring the three-dimensional shape of a component and its alignment by a fast scan by some light source has been verified, but is not as yet of sufficient accuracy for most applications. Its successful development would revolutionize setting-up procedures.

The problem of in-process measurement of component size is moving towards certain solutions. Touch trigger probes have been suggested, as have optical devices, each being claimed to make in-process measurement. There is some controversy in this area as the cutting force, coolant, swarf, heat distortion, and smoke hinder the analysis by probes and optical systems.

The removal of swarf and smoke can possibly be achieved by using compressed air, but the removal of coolant is more difficult, and the immediate removal of heat distortion appears impossible. Laser-based methods are non-contact and these and other optical methods offer a good solution to in-process measurement, since they have large stand-off distances. If there is little problem of swarf and coolant, then probes and other proximity sensors are extremely useful and very reliable in performing the required tasks.

Cutting forces and heat distortion will not have the same effect and importance with different components; the problem is component-orientated. If the component has close tolerances, a solution may be to use an in-cycle measuring station. This would be used post-process, perhaps after washing. A disadvantage of this method would be the possible loss of the first component of the batch. This again indicates a component-orientated problem, suggesting that costly components should be measured in-process.

Surface finish offers good scope for in-process measurement, with optically based methods most likely to succeed; for example, using image analysis.

All the sensors discussed in this report have their own particular application, and some may be used for multiple applications. It is essential when looking at a machining system to analyse the system thoroughly, identifying exactly which parameters need to be ana-lysed, and to what accuracy and magnitude they are to be monitored. Areas such as trend analysis should also be considered for the possible introduction as maintenance methods (predictive maintenance). Through the introduction of maintenance techniques to condition monitoring a better picture can be gained of changes brought about by the introduction of the monitoring systems and, therefore, the pay-back period.

6.2 CONDITION MONITORING

6.2.1 Abstract

The primary objective of a condition-based monitoring system, such as predictive maintenance, is to prevent unscheduled down-time. This is done by monitoring the parameters which are related to the condition of the machine. Corrective action can then be properly planned and down-time can be minimized. The move by industry towards using advanced manufacturing systems, which run almost at

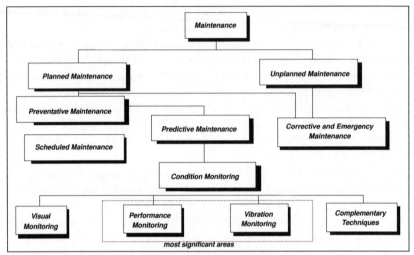

Fig. 6.1 Maintenance management system

full system capacity, and the use of production techniques, such as Just-In-Time (JIT), has made unscheduled down-time even more undesirable. Sensible use of condition monitoring can reduce maintenance costs.

As quoted by Harries (1987):

The CBM (condition based monitoring) methodology, when properly applied, can result in optimised labour and material expenditure for an effective maintenance service.

6.2.2 Benefits

The benefits to the user of applying condition monitoring to a machining system are improved system availability and the reduction of scrap. Automatic monitoring of critical operations is essential for unmanned machining. Automatic diagnosis, which identifies the cause of any failure, provides a tool to reduce down-time to a minimum. Also, automatic prediction of failures enables corrective maintenance to be planned into the production schedule.

Hence, condition-based monitoring offers an alternative way to combat unscheduled down-time, in that it allows the equipment to be operated until a breakdown is imminent.

To ensure optimum performance and pay-back, each application of predictive maintenance is specifically designed and tailored to suit the particular requirements of the situation.

Figures from Pinnacle Technical Management Services Ltd in Harlow, Essex (UK), show some typical savings expected in an average British manufacturing company:

- 40 per cent reduction in plant downtime;
- 20 per cent reduction in maintenance costs;
- 15 per cent reduction in spares costs;
- 10 per cent reduction in manpower costs.

From these figures we can see that there may be large benefits from following a predictive maintenance plan.

To obtain maximum benefit from the adoption of predictive maintenance techniques, they must be properly be integrated with the plant management function, with due consideration to:

- operational suitability;
- methods of application;
- degree of staff training required;
- intrinsic value of information;
- how the information can be used;
- maximum return on the investment.

These are achieved through management modules, designed in building-block form (see Fig. 6.2). Their progressive introduction into the operation of both the commercial and human elements of the organization allows predictive maintenance to achieve full effect.

Implementation

The implementation of predictive maintenance can only have full effect, especially in economic terms, with a total commitment by staff and through a clear understanding of the philosophy behind its implementation. It is vital that an integrated approach is established at an early stage.

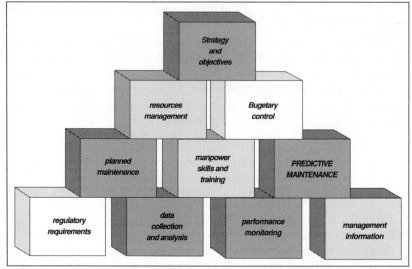

Fig. 6.2   Management modules for predictive maintenance

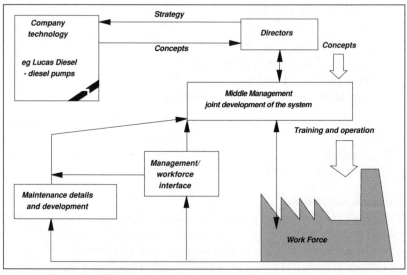

Fig. 6.3   Maintenance integration within an organization

Figure 6.3 illustrates graphically these concepts. Staff must be properly trained in the use of such management tools, and must be made aware of the importance of the high organizational content inherent in them. Condition monitoring systems can provide the engineer and manager with important information about the reliability of the plant, but the technology must be properly applied and administered.

6.2.3.1 Changes incurred through implementation

If a predictive maintenance scheme is introduced into an organization it creates many changes throughout the plant, not just for the maintenance staff.

Stock control
The spare parts for the machines would have to be analysed very closely. Currently there may be many parts which are kept in storage which are very rarely used, or which can be very easily obtained quickly. These parts take up valuable factory space, and money is wasted in purchasing and holding them which could be put to other uses of more importance. However, parts which are regularly replaced or parts which are difficult to obtain and would cause catastrophic effects if they were to fail must be kept in storage.

Through preventative maintenance, if a component of a machining system is found to be deteriorating, sufficient notice may be given to order a replacement component, without the necessity of holding that component in storage.

Purchasing
In conjunction with the spare parts stores, purchasing may routinely order particular components, for example, a standard thrust bearing ordered once every two months. If parts are ordered with the assumption of a part failing, without any evidence, then these parts may become obsolete. Conversely, parts may be found to fail within a certain period, and purchasing to a set schedule may prove to be beneficial. However, purchasers must be made aware that if parts *are* ordered, then they may be required urgently.

Scheduling

In the past, with breakdown maintenance, scheduled work had to take into account a random variable – that one of the machines my become unserviceable. Later, with the transition to planned maintenance the work schedule became more predictable but still allowed for the random breakdown of a machine. With predictive maintenance a breakdown should be determined before the occurrence, so reducing the random element significantly. With a predictive maintenance programme, if a machine is diagnosed as developing a fault, then the schedule must adapt to the change, building around the fact that one machine will be out of use. With this programme, time is allowed for this adaptation of the work schedule.

This can be controlled by a link between the machining systems and the main controlling/scheduling computer. If a machine if found to be developing a fault then the computer recognizes this and the necessary adjustments are made automatically to the schedule, with the minimum disturbance and disruption to the work.

This system in turn is linked with the raw material supplies, ensuring the correct flow; this may be amalgamated with such systems as JIT and *kan-ban*.

Personnel

Employees directly involved with the maintenance of the machining systems will require full training with the apparatus for predictive maintenance, together with a full understanding of condition monitoring systems. Persons indirectly involved with the system, such as stores workers, should also be familiar with the features of the maintenance techniques, enabling a quick and efficient service from wherever it may be required throughout the organization.

Financial

After determining the need for preventive maintenance and the critical machining systems that require this technique, the financial justification for introducing such a system must be established. There are many systems on the market, each with differing capital costs, training costs, and other service costs. Some companies specialize in package deals involving site feasibility studies, introducing pilot schemes, implementing the predictive system, and on-site training courses. The exact requirements of the organization must be determined and the most economic way of achieving them determined. Once a cost has been determined then it must be justified. This may be done in several ways:

- accounting rate of return;
- minimum payback period;
- net present value;
- internal rate of return.

This most popular of these methods is the minimum payback period, as this is most easily understood, and calculated, but with any of these methods the results are the most important area in introducing the system.

6.3 SUMMARY

In reality, predictive maintenance may be used in conjunction with planned maintenance and, infrequently, breakdown maintenance will be required. The degree to which each type of maintenance is introduced results in the required end product. Some organizations require a mixture of maintenance techniques as some machining systems are so reliable that it is only very infrequently that a fault occurs. Alternatively, a machine may be vary rarely used, and applying predictive maintenance to these areas would result in a very long payback period (see section 5.4).

The description of the various types of sensors given in the first four chapters of this book should give a sufficient knowledge of the areas of a machining system which are able to be monitored, and with what type of sensor or sensory system this can be achieved.

Literature

BIBLIOGRAPHY

BABANI, B. *Practical electronic sensors* (1981) Bishop Owen.

BILLINGTON, R. and ALLAN, R.N. *Reliability evaluation of engineering systems: concepts / techniques* (1991) Plenum Press, New York.

BROZEL, M.R. and STINLAND, J.J. *Defect recognition in semiconductors before and after processing* (1991) Adam Hilger, Bristol.

DIMAROGONAS, A.D. *Reliability data collection and analysis* (1991) Prentice Hall, Englewood Cliffs, NJ.

GOPEL, W. and HESSE, J. *Optical sensors* (vol. 6) VCH, London.

GRATTAN, K.T.V. *Sensors* (1991) Adam Hilger, Bristol.

GREEN, W.B. *Digital image processing: a systematic approach* (1990) Van Nostrand Reinhold, New York.

MINKOFF, J. *Signals, noise and active sensors: radar, sonic, laser / radar* (1991) John Wiley, New York.

MIRZAI, A.R. *Artificial intelligence: concepts and applications in engineering* (1990) Chapman & Hall, London.

MYERS, D.J. *Digital signal processing* (1990) Prentice Hall, Englewood Cliffs, NJ.

SATYAM, M. and RAMKUMAR, K. *Foundations of electronic devices* (1990) John Wiley, New York.

SINCLAIR, I.R. *Sensors and transducers: a guide for technicians* (1992) Newnes Butterworths, Sevenoaks.

SPC Statistical Process Control (1990) Ford Motor Company.

SKVOR, Z. *Vibration systems and their equivalent circuits* (1991) Elsevier Science Publishing, Oxford.

121

USHER, M.J. *Sensors and transducers* (1985) Macmillan Press, London.

REFERENCES

DRAKE, P.R. 'Condition monitoring', University of Wales Institute of Science and Technology.

MAYER, J.E. 'Cutting tool sensing', Conference on Sensors, Detroit, MI (MS86-1004).

Index